SANITARY LANDFILL LEACHATE

SANITARY LANDFILL LEACHATE

Generation, Control and Treatment

SYED R. QASIM

Professor of Civil Engineering
The University of Texas at Arlington
Arlington, Texas

WALTER CHIANG

President
Chiang, Patel and Associates, Inc.
Consulting Engineers
Dallas, Texas

TECHNOMIC
PUBLISHING CO., INC.
LANCASTER · BASEL

Sanitary Landfill Leachate

a **TECHNOMIC**⸰publication

Published in the Western Hemisphere by
Technomic Publishing Company, Inc.
851 New Holland Avenue, Box 3535
Lancaster, Pennsylvania 17604 U.S.A.

Distributed in the Rest of the World by
Technomic Publishing AG
Missionsstrasse 44
CH-4055 Basel, Switzerland

Printed in the United States of America

10 9 8 7 6 5 4 3 2

Main entry under title:
  Sanitary Landfill Leachate—Generation, Control and Treatment

A Technomic Publishing Company book
Bibliography: p.
Includes index p. 323

Library of Congress Catalog Card No. 94-60645
ISBN No. 1-56676-129-8

Table of Contents

Preface

SANITARY landfills are the most widely utilized method of solid waste disposal around the world. With increased use and public awareness of this method of disposal, there is much concern with respect to the pollution potential of the landfill leachate. Depending on the composition and extent of decomposition of the refuse and hydrological factors, the leachate may become highly contaminated. As leachate migrates away from a landfill, it may cause serious pollution to the groundwater aquifer as well as adjacent surface waters. There is growing concern about surface and groundwater pollution from leachate. Better understanding and prediction of leachate generation, containment, and treatment are needed.

This book contains a literature review of various methodologies that have been developed for prediction, generation, characterization, containment, control, and treatment of leachate from sanitary landfills. The contents of this book are divided into nine chapters. Each chapter contains theory and definition of the important design parameters, literature review, example calculations, and references.

Chapter 1 is devoted to basic facts of solid waste problems current status, and future trends towards waste reduction and recycling. Chapter 2 provides a general overview of municipal solid waste generation, collection, transport, resource recovery and reuse, and disposal options. The current status of sanitary landfill design and operation, problems associated with the landfilling and future trends are presented in Chapter 3. Methods of enhanced stabilization, recycling landfill space, methane recovery, and above grade landfilling, and closure and post closure care of completed landfills, are also discussed in detail. Chapter 4 provides a general overview of *Subtitle D* regulations, and its impact upon sanitary landfilling practices. Chapter 5 is devoted entirely to moisture routing and leachate generation mechanisms. Examples of calculation procedure for determining the leachate quantity produced at a landfill are presented. Chapter 6 is devoted to chemical characterization of leachate that changes over the life of the fill. Both theoretical and experimental results are provided to estimate the leachate quality. Chapter 7 provides leachate attenuation processes and mechanisms. Chapter 8 is devoted to leachate collection systems. Natural soil sealants, admixed materials and synthetic membranes, their effectiveness, and methods of installation and economics are fully discussed. Chapter 9 provides a detailed review of leachate treatment methodology. Kinetic coefficients and treatment plant design considerations are summarized for the sole purpose of assisting consultants to design leachate treatment facilities. Leachate treatment case histories, and numerous process trains are presented for treating leachate from young landfills, and how the process train can be changed effectively as leachate quality changes with time.

This book will serve the growing needs of consulting engineers, professionals, equipment supplies, and technical personnel in city, state, and federal organizations who must design and review the design and specifications of sanitary landfills.

This book is a joint effort to utilize academic and consulting engineering resources. We believe that this joint project provides a unique combination of academic and consulting engineering experience in development of a more complete document on landfill leachate generation and control.

SYED R. QASIM
WALTER CHIANG

Acknowledgements

W E are grateful to many individuals and organizations who have assisted us with the preparation of this book. Max Spindler of the University of Texas at Arlington reviewed Chapter 7. George Yazdani of Poly-Flex, Inc. reviewed portions of Chapter 3, and entire Chapter 8 and provided valuable suggestions. Poly-Flex also supplied many photographs and figures for use in the book. Dr. P. R. Senthilnathan of Zenon Environmental Inc. reviewed Chapter 9, and provided many constructive suggestions. He also supplied numerous technical reports and articles.

Many students assisted with this book. Rex Miller diligently conducted library search to obtain copies of reports and journal articles and proofread the manuscript. Rex Miller and Shih Pan prepared the indexes. Many students also reviewed several chapters. Saidara Waheed and Beverlee Moore typed portions of the manuscript.

Chiang, Patel and Associates, Inc., a consulting engineering firm of Dallas, Texas, provided technical support. Most of the artwork was prepared by them. Many professionals from this organization reviewed portions of several chapters.

Finally, we must acknowledge with deep appreciation the support and encouragement of our families.

CHAPTER 1

Sanitary Landfill Leachate: Generation, Control and Treatment

1.1 INTRODUCTION

SOLID waste has finally received some respect, or at least some serious attention in recent years. U.S. municipalities are generating solid wastes at record quantities, over 200 million tons* per year; and solid waste collection and disposal constitutes the fastest growing segment of the municipal budget. Besides money, much more is at stake. Environmental pollution, water quality deterioration, space limitations, landfills reaching full capacity, public opposition hindering new site development, loss of valuable resources, and many more issues are being raised by citizens, legislators, regulators, and public officials.

Perhaps 1992 has seen an accelerated interest in solid waste reduction and recycling activity at state and local levels. Almost all states have now instituted, initiated or seriously evaluated some form of solid waste management plan with schedule for reducing waste volumes. Some states have waste reduction goals, some have recycling goals, some have goals requiring both waste reduction and recycling; some goals are mandatory, some are voluntary. Local governments are also expected to develop and implement a plan to achieve the waste reduction guidelines.

1.2 INTEGRATED SOLID WASTE MANAGEMENT

Solid waste generation is a fact of modern living. Everyone contributes to the problem. Most residents take solid waste collection and disposal at low cost for granted. Some are interested in recycling programs to lower costs and conserve resources. Public interest groups would like to see

---

*1 ton = 907.2 kg (0.9 metric ton).

stricter legislation to encourage citizens to recycle, and pressure manufacturers for less wasteful product packaging. Some groups advocate resource conservation and environmental protection at any cost. Local governments are interested in managing solid waste collection and disposal in an efficient, economical, and environmentally acceptable manner with least controversy. Waste handlers and landfill operators are concerned with an operation that meets the minimum standards, minimizes cost and public liability, and maximizes workers' safety. Equipment manufacturers want to provide their customers with the best equipment for solid waste collection, disposal, and resource recovery, and maximize their profit. Resource recycling industries are concerned with market fluctuations, quality of secondary material, and long-term commitments and financial arrangements that may be needed for process modifications. Many of these interests are conflicting, and reconciling these special interests can make development of an effective solid waste management plan very complex and controversial. Knight (1988) stated that the problem with solid waste is not that there is too little, but too much for everyone. At this time, however, there is a great interest in resource recycling, and energy recovery to reduce wastes and to extend the life of landfills for final disposal. Future landfills will be more carefully regulated by strict design, construction, and operation criteria, and long-term post-closure care.

The Office of Solid Waste published the *Solid Waste Dilemma: An Agenda for Action* (U.S. EPA 1988). The Agenda lays out EPA's strategy for dealing with municipal solid waste. It identifies the roles of the many players in solving the waste management problems, and the specific activities EPA is committed to conducting. EPA also published the *Decision Maker's Guide* to set an agenda for solving the nation's garbage problems (U.S. EPA 1989).

As explained in these publications, EPA encourages municipalities to use a mix of solutions to handle waste, since there is no single management approach that can effectively address all issues of the solid waste problem. Within the range of management options, EPA suggests a four level hierarchy for decision makers to consider when planning and implementing integrated waste management. The first level of the hierarchy is *source reduction*. Individuals, government, commercial establishments and industries are expected to participate in source reduction by reducing the quantity of solid, and toxic wastes.

The second level of the hierarchy is *recycling*. This involves collecting, reprocessing, marketing, and using materials that were once considered trash. Many of the components of our waste stream can be recycled—from metals and plastics to used oil and yard waste. The third level in EPA's hierarchy is waste *combustion*. Combustion can be used to reduce the volume of the waste stream and to recover energy. Finally, the fourth level of hierarchy is *landfilling*. Landfilling is the only true disposal option. It is a necessary component of waste management, since all management options produce some residue that must be disposed of through landfilling. EPA, and State and local governments are working hard to improve the safety of both combustion and landfilling, through new regulatory controls, design and operational practices, training, and careful monitoring.

1.3 RECYCLING AND WASTE REDUCTION

Many state and local legislative actions have been taken since the mid-1980s to encourage recycling of various components of municipal solid wastes. Miller (1993) provided a 1992 update on recycling status in the United States. Currently, 39 states and the District of Columbia have some form of statewide recycling law, although the requirements vary greatly. Twenty states require that local governments prepare recycling plans in order to meet waste reduction or recycling goals. In many instances, the "recycling plan" does not explicitly require recycling, but waste reduction is required. The remaining 12 states require local governments to ensure that recycling opportunities are available, either through curbside collection, drop-off centers, or mechanical processing of mixed solid wastes.

1.3.1 STATE GOALS AND WASTE REDUCTION

The recycling goals and waste reduction plans, and target years to achieve these, vary greatly. Miller (1993) provided a summary of waste recycling, and reduction status of 40 states and the District of Columbia (Table 1.1). In these plans the definition of recycling and waste reduction also vary greatly. As an example, in some instances composting and/or energy recovery are considered recycling. Likewise, waste reduction may be achieved by total ban, or minimum content requirements. Examples of total ban and minimum content requirements are lead-acid and mercury oxide batteries, mercury containing products such as fluorescent lights, motor oil and other hazardous wastes, whole tires, glass, metal containers, rigid plastics, yard wastes, corrugated boxes, telephone directories, and materials with newspapers. Miller (1993) provided the minimum-content laws for 13 states (Table 1.2).

1.3.2 MARKET DEVELOPMENT AND INCENTIVES

Many states have instituted tax credits to support recycling and waste reductions. Some states have even encouraged statewide market development programs for secondary material. Most incentives come in the form of in-

TABLE 1.1 **State Goals of Municipal Solid Waste Recycling, Reduction and Diversion.**

State	Goal	Year
Alabama	25%	
Arkansas	40%	2000
California	50%	2000
Connecticut	25%	1991
D.C.	45%	1996
Florida	30%	1994
Georgia	25%	1996
Hawaii	50%	2000
Illinois	25%	2000
Indiana	50%	2001
Iowa	50%	2000
Kentucky	25%	1997
Louisiana	25%	1992
Maine	50%	1994
Maryland	20%	1994
Massachusetts	56%	2000
Michigan	40–60%	2005
Minnesota	25%	1993
Mississippi	25%	1996
Missouri	40%	1998
Montana	25%	1996
Nebraska	50%	2002
Nevada	25%	
New Hampshire	40%	2000
New Jersey	60%	1995
New Mexico	50%	2000
New York	50%	1997
North Carolina	25%	1993
North Dakota	40%	
Ohio	25%	1994
Oregon	50%	2000
Pennsylvania	25%	1997
Rhode Island	70%	
South Carolina	30%	1997
South Dakota	50%	2001
Tennessee	25%	1995
Texas	40%	1994
Vermont	40%	2000
Virginia	25%	1995
Washington	50%	1995
West Virginia	50%	2010

Source: Adapted from Miller (1993) with permission from *Waste Age*.

TABLE 1.2 Municipal Solid Wastes Minimum Content Laws of Several States.

State	Waste Reduction, and Year
Arizona	Newsprint: 50% by 2000
California	Newsprint: 50% by 2000 Glass containers: 65% by 2005 Plastic bags: 10% for 1.0 mil bags, 1993; 30% for 0.75 mil bags, 1995
Connecticut	Newsprint: 50% by 2000 Telephone directories: 40% by 2001
Florida	Newsprint: taxed 50 cents per ton if fail to meet 50% recycled content in 1992
Illinois	Newsprint: 23% by 1993
Maryland	Newsprint: 40% by 1998 Telephone directories: 40% by 2000
Missouri	Newsprint: 50% by 2000
North Carolina	Newsprint: 40% by 1997
Oregon	Glass containers: 50% by 2000 Newsprint: 7.5% by 1995 Plastic containers: 25% post-consumer material or 25% recycling rate or reusable/refillable 5 times or reduced 10% in content by 1995 (certain exemptions)
Rhode Island	Newsprint: 40% by 2001
Texas	Newsprint: 30% by 2000
West Virginia	Newsprint: "highest practicable content," advisory committee to determine rate
Wisconsin	Newsprint: 45% by 2001 Plastic containers: 10% by 1995

Source: Adapted from Miller (1993) with permission from *Waste Age.*

come tax credits; corporate tax credits for purchases of equipment; credits for purchase of secondary material; sales tax on recycling equipment; property tax exemption for buildings, equipment, and land involved in recycling products; and many more types of tax and credit incentives. Miller (1993), provided a summary of such credits for 27 states. Specific information on tax credits and incentive programs may be obtained by contacting the State Commerce, Economic Development, or Tax Offices.

1.4 ENERGY RECOVERY AND WASTE REDUCTION

Incineration of municipal solid wastes has been practiced to reduce waste volume, and recover energy. Again, the prime motivation is due to increasing scarcity of landfill space, combined with the rising cost of trucking wastes to more distant landfills. Many communities have built refuse-to-energy plants that are fast replacing landfill. There are currently 190 municipal waste combuster plants in the USA that exceed the total processing capacity of 31 million tons of MSW per year. These facilities do not include refuse derived fuel (RDF) processing plants, or simple incinerators (Kiser 1992). Many more waste-to-energy combusters are planned or under construction. The Greater Detroit Resource Recovery Plant, the country's largest plant has a capacity of 4000 tpd that will generate 250,000 kg (550,000 lbs) of steam per hour (Nasenbeny 1987). Wheelabrator Technologies, based in Danvers, Massachusetts is operating 8 facilities handling 4.5 million tons of solid wastes annually and produces 325 MW power daily (Knight 1988). Despite high capital outlay for mass-burn facilities, the option is perceived as economically competitive in many communities, particularly where landfill space is scarce or nonexistent.

1.5 LANDFILLING AND LEACHATE CONTROL

In the United States, sanitary landfills have been the most popular method of municipal solid waste disposal. It was estimated that about 6,500 solid waste landfills existed prior to 1988, when Subtitle D of the Resource Conservation and Recovery Act (RCRA–U.S. EPA 1991) set forth new rules governing the design, construction and operation of new landfills, and the closure of old fills. As the Subtitle D regulations take effect on October 9, 1993, many landfills will close because of space limitations or noncompliance. By the year 2000, EPA estimated that in spite of increased recycling, waste reduction, and incineration, approximately 49 percent of the municipal solid waste will still be landfilled. This means that almost 82 percent of the current landfill capacity will still be needed by the end of the century (Repa and Sheets 1992). EPA estimates suggest that about 2100 landfills will be active by the end of the century.

Under Subtitle D regulations proper leachate management must be implemented. The landfills must be properly lined so that leachate does not seep into the groundwater. The collected leachate must be treated. Leachate is classified as an industrial waste under federal pre-treatment guidelines. This means that any publicly owned treatment works (POTW) treating leachate must prevent it from interfering with the treatment process, or harming its sludge quality.

Under Subtitle D, the guidelines for landfill leachate treatment and disposal are established by states. Additional leachate management guidelines may be developed in the future. Proposed regulations–40 CFR, Subchapter N, "Central Waste Treatment Facilities Commercial Pretreatment

Effluent Guidelines/Standards"—could take effect as early as 1995 and further define liabilities associated with off-site disposal of leachate from landfills (Copa 1992).

1.6 FUTURE TRENDS

It is expected that in future, municipal solid waste management practices will greatly emphasize in resource recovery, and solid waste reduction. Incineration with energy recovery will play an important role in waste reduction and energy conversion. Sanitary landfills, however, will continue to be used for solid waste and residue disposal. The design of landfill, leachate treatment, and closure requirements will continue to be the major issues to protect our groundwater resources. The major thrust of this publication is to address the issues dealing with leachate containment and treatment, and protection of groundwater resources.

1.7 REFERENCES

Copa, W. M. 1992. "Landfill Leachate Presents a Challenge," *Environmental Protection,* December:19–22.

Kiser, J. V. L. 1992. "Municipal Waste Combustion in North America: 1992 Update," *Waste Age,* 23(11):26–34.

Knight, A. 1988. "Energy From Waste: Recovering a Throwaway Resource," *EPRI Journal,* October/November:25–35.

Miller, C. 1993. "Recycling in the State, 1992 Update," *Waste Age,* 24(3):26–34.

Nasenbeny, R. J. 1987. "Refuse-to-Energy Plant Gets City Out of Dumps," Consulting Specifying Engineer, March:54–56.

Repa, E. W. and S. K. Sheets. 1992. "Subtitle D's Effects on Private Industry," *Waste Age,* 23(4):267–272.

U.S. Environmental Protection Agency. 1988. *The Solid Waste Dilemma: An Agenda for Action-Background Document.* EPA/530-SW-88-054A, U.S. EPA Office of Solid Waste, Washington, D.C.

U.S. Environmental Protection Agency. 1989. *Decision-Makers Guide to Solid Waste Management.* EPA/530-SW89-072, Washington, D.C.

U.S. Environmental Protection Agency. 1991. "Solid Waste Disposal Facility Criteria; Final Rule," *40 CFR Part II, 257 and 258,* Subtitle D of the Resource Conservation and Recovery Act (RCRA).

CHAPTER 2

Municipal Solid Waste Management – An Overview

2.1 INTRODUCTION

SOLID wastes may be defined as useless, unused, unwanted or discarded material available in solid form. Semisolid food wastes and municipal sludge may also be included in municipal solid waste. The subject of solid wastes came to the national limelight after the passage of the *Solid Waste Disposal Act of 1965* (U.S. Congress 1965). Today, solid waste is accepted as a major problem of our society.

In the United States over 180 million tons of municipal solid waste (MSW) was generated in 1988 (U.S. EPA 1990). At this generation quantity, the average resident of an urban community is responsible for more than 1.8 kg (4.0 lbs) of solid waste per day. This quantity does not include industrial, mining, agricultural, and animal wastes generated in the country each year. If these quantities are added, the solid waste production rate reaches 45 kg per capita per day (100 lb/c.d.). To introduce the reader to the solid waste management field, an overview of municipal solid waste problems, storage, collection, resource recovery, and disposal methods are presented in this chapter. Since leachate generation and control is the major thrust of this book, greater emphasis has been given to the design and operation of municipal sanitary landfills, regulations governing land disposal, and leachate generation, containment and treatment methods. These topics are covered in greater depth in separate chapters.

2.2 MUNICIPAL SOLID WASTES QUANTITIES

Municipal solid waste (MSW) includes wastes such as durable goods, nondurable goods, containers and packaging, food wastes, yard wastes, and miscellaneous inorganic wastes from residential, commercial, institutional, and industrial sources. Examples of waste from these categories include appliances, newspapers, clothing, food scraps, boxes, disposable tableware, office and classroom paper, wood pallets, and cafeteria wastes. MSW does not include wastes from sources such as municipal sludges, combustion ash, and industrial nonhazardous process wastes that might also be disposed of in municipal waste landfills or incinerators.

Determining actual MSW generation rates is difficult. Different studies report a wide variation as they use different components. Many times industrial and demolition wastes are included in municipal solid wastes. The U.S. EPA (1990) has estimated that a total of over 180 million tons of MSW was generated in the United States in 1988, and that generation is rising at a rate of slightly over 1 percent each year. This estimate is based on a material flow model utilized by Franklin Associates, and is generally referred to as the EPA/Franklin model (U.S. EPA 1988, 1990). Most of the increase in the MSW generation rate is due to population growth. How-

TABLE 2.1. Various Components of Solid Wastes Generated in MSW.

	Weight Generated (Millions of Tons)	Weight Component, %
Paper and Paperboard	71.8	40.0
Glass	12.5	7.0
Metals		
Ferrous	11.7	6.5
Aluminum	2.5	1.4
Other Nonferrous	1.1	0.6
Total Metals	15.3	8.5
Plastics	14.4	8.0
Rubber and Leather	4.6	2.6
Textiles	3.9	2.2
Wood	6.5	3.6
Other	3.1	1.7
Total Nonfood Product Wastes	132.1	73.6
Other Wastes		
Food Wastes	13.2	7.3
Yard Wastes	31.6	17.6
Miscellaneous Inorganic Wastes	2.7	1.5
Total Other Wastes ·	47.5	26.5
Total MSW	179.6	100.0

1 Ton = 2000 lbs, 907.2 kg (0.9 metric ton).
Source: U.S. EPA (1990).

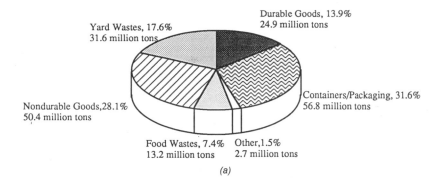

Yard Wastes, 17.6%
31.6 million tons

Durable Goods, 13.9%
24.9 million tons

Nondurable Goods,28.1%
50.4 million tons

Containers/Packaging, 31.6%
56.8 million tons

Food Wastes, 7.4%   Other,1.5%
13.2 million tons   2.7 million tons

(a)

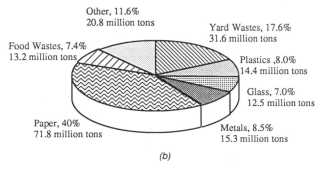

Other, 11.6%
20.8 million tons

Food Wastes, 7.4%
13.2 million tons

Yard Wastes, 17.6%
31.6 million tons

Plastics ,8.0%
14.4 million tons

Glass, 7.0%
12.5 million tons

Paper, 40%
71.8 million tons

Metals, 8.5%
15.3 million tons

(b)

Figure 2.1 Estimated portions of products and materials in MSW, 1988, by weight. (a) Products by weight. Total weight 179.6 million tons; (b) Materials generated by weight. Total weight 179.6 million tons. Source: U.S. EPA (1990).

ever, each person is also generating more waste on an average. It is estimated that, without source reduction, MSW generation will increase to 200 million tons or 1.9 kg/c·d (4.2 lb/c·d) by the year 1995, and 216 million tons or 2.0 kg/c·d (4.4 lb/c·d) by the year 2000. Based on current trends and information, EPA projects that 20 to 28 percent of MSW will be recovered annually by 1995. Exceeding this projected range will require fundamental changes in government programs, technology, and corporate and consumer behavior (U.S. EPA 1989, 1990). The composition of MSW by weight in terms of products estimated by EPA is provided in Table 2.1 and Figure 2.1. The products category [Figure 2.1(a)] includes durable goods such as major appliances, furniture, rubber tires and miscellaneous. Nondurables are newspapers, books, magazines, tissue paper, office and commercial paper, clothing, footwear, and miscellaneous. The broad categories of materials in MSW are made up of many individual products. The weight fraction and percent components of these materials are given in Figure 2.1(b). The breakdown of quantities of waste going to recycling, combustion, and landfilling are shown in Figure 2.2(a). The recovery of materials for recycling and composting was estimated at 13 percent in 1988. Approximately 14 percent MSW was incinerated with some energy and resource recovery. Over 73 percent of MSW is disposed of in sanitary landfills. The volume occupied in sanitary landfills in 1988 was over 306 million m³ (400 million yd³). The volume and percent of different components in landfills are given in Figure 2.2(b).

2.3 OTHER SOLID WASTE SOURCES

Large quantities of solid wastes are also generated from many sources other than MSW. These quantities, although generated away from urban areas, or collected separately from municipal wastes, are enormous. These solid waste sources are industrial, agricultural, mining and animal wastes. These quantities, as reported by Tchobanoglous et al. (1977), and

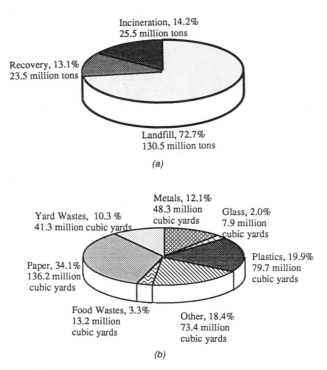

Figure 2.2 1988 solid waste management practices in U.S.A. (a) Disposal practices, total weight 179.6 million tons; and (b) Volume of discards in sanitary landfills, total volume 400 million cubic yards. Source: U.S. EPA (1990).

TABLE 2.2. Solid Waste Quantities Other Than MSW Generated in the United States.

Solid Waste Source	Generated Quantity 10^6 tons/year
Industrial	140
Agricultural	640
Animal	1700
Mining	1700
Total	4180

1 ton = 907.2 kg.
Source: Adapted from Tchoganoglous et al. (1977), and Henry and Heinke (1989).

Henry and Heinke (1989) are provided in Table 2.2. The industrial wastes are generally particular to the individual industry. They may include scrap metals, lumber, plastics, bales of waste paper, boards and rags, used-up drums and containers, slags, sludges, etc. The agricultural wastes consist of residues from most crop harvestings, horticulture wastes and orchard prunings.

The animal wastes are generated from feedlots, dairies, chicken ranches, pig factories and other confined animal raising operations. The waste is generally manure, bedding material, punch manure from slaughterhouses, and carcasses. The major components of mining wastes are the by-products of mining operations and processing of minerals and fossil fuels. Increased waste production is anticipated from all sources, not because of greater production, but because of use of lower grade ores.

2.4 HAZARDOUS WASTES

Hazardous wastes include explosive, flammable, volatile, radioactive, toxic, and pathological wastes. These wastes require special care in storage, collection, transportation, treatment, and disposal to prevent harm to human health, animals, and property. These wastes can increase serious irreversible, or incapacitating illness, or pose potential hazards to human health or environment when improperly managed. In 1992, the U.S. Environmental Protection Agency published a complete revised list, categorization, and specific definition of hazardous wastes (U.S. EPA 1992). The list of hazardous wastes in the EPA document is very extensive, and classification is complex. In general, this list includes spent halogenated solvents for degreasing (trichloroethylene, methylene chloride, and others), spent nonhalogenated solvents (xylene, ethyl benzene, ethyl ether, acetone, and others), electroplating baths, wastes from distillation processes, wastewater sludges from many industrial production processes, specific substances identified as acute hazardous wastes (silver cyanides, toxaphene, arsenic oxide, and others), specific substances identified as hazardous wastes, and wastes that exhibit properties of *ignitability, corrosivity, reactivity,* or *toxicity.* Examples of wastes exempted from hazardous waste lists are municipal solid wastes and wastewater treatment plant sludges, irrigation return flows, mine tailing, animal manure, fly ash, drilling fluids, and wastes from crude oil, natural gas, or geothermal energy development. Nuclear and other radioactive wastes are controlled under separate regulations.

The new Subtitle D rules impose several general design and operation criteria for MSW landfills (U.S. EPA 1991). One of the requirements is that landfill operators develop a program to detect and prevent disposal of regulated hazardous wastes into municipal landfills. The program will involve at a minimum: (1) random inspection of incoming loads, (2) inspecting suspicious loads, (3) maintaining records of inspections, (4) training personnel to locate hazardous waste in the general municipal solid waste stream, and (5) enacting reporting procedures when such wastes are found.

Since the thrust of this book is municipal solid wastes, and in particular leachate generation, characterization, containment and disposal from municipal sanitary landfills, no additional details on hazardous solid waste are provided in this book. Readers are referred to several excellent sources on this subject (Wentz 1989; U.S. EPA 1989; Davis and Cornwell 1991).

2.5 MUNICIPAL SOLID WASTE MANAGEMENT

A successful solid waste management system utilizes many functional elements associated with generation, on-site storage, collection, transfer, transport, characterization and processing, resource recovery and final disposal (U.S. EPA 1976). All these elements are interrelated, and must be studied and evaluated carefully before any solid waste management system can be adapted. It is a multidisciplinary activity involving engineering principles, economics, and urban and regional planning. Figure 2.3 illustrates a flow diagram encompassing these functional elements and decision points that must be made from the generation point to ultimate disposal. The material presented in this chapter in general follows these functional elements.

2.5.1 SOURCES AND CHARACTERISTICS

Municipal solid waste (MSW) or urban solid waste is normally comprised of food wastes, rubbish, demolition and construction wastes, street

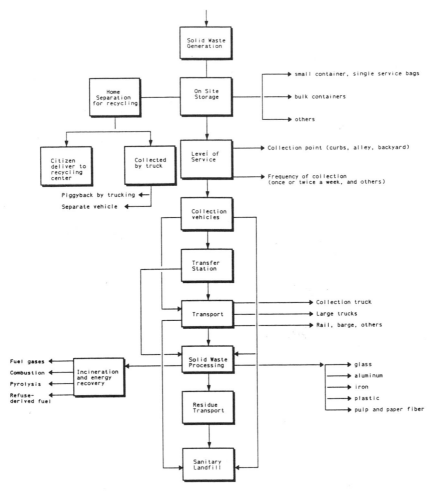

Figure 2.3 Functional elements and decision points of a solid waste management system. Source: U.S. EPA (1976).

sweepings, garden wastes, abandoned vehicles and appliances, and treatment plant residues. Definitions of the various components of MSW are given in Table 2.3. Quantity and composition of MSW vary greatly for different municipalities and time of the year. Factors influencing the characteristics of MSW are climate, social customs, per capita income, and degree of urbanization and industrialization. The typical categories and sources of MSW as reported in the literature are summarized in Table 2.4.

The composition of MSW as collected may vary greatly depending upon geographical region and season. The range and typical values of various components and their moisture contents are summarized in Table 2.5. The

TABLE 2.3. **Classification Based Upon Character of Material.**

Kind	Definition and Composition	Source
Garbage	Animal and vegetable product composed of putrescible organic matter. Contains enough food value that it can be used as commercial animal feeds. It has high grease content and can be converted into compost.	Home and restaurant kitchen and food servings; supermarkets; and meat, poultry, fish, vegetable and fruit markets.
Combustible rubbish— synonymous with trash	Combustibles and noncombustibles (does not include garbage). Combustible rubbish consists mainly of the organic component of refuse such as rags, paper, boxes, wood, plastics, lawn trimmings, etc.	Homes, stores and institutions.
Noncombustible	Noncombustible rubbish is bottles, cans, dust, metals, etc.	Homes, stores and institutions.
Street refuse	Solid wastes collected from the city roads and sidewalks. It contains dirt, paper, leaves, catchbasin dirt, dead animals. The waste contains combustible and noncombustible rubbish.	Municipal services such as streets, sidewalks, landscaping, and other municipal property.
Demolition wastes	Lumber, pipes, bricks, stores, masonry, and other construction materials. The waste is generally of a noncombustible nature.	Building demolition sites, and road repairs and excavation.
Construction wastes	Scrap lumber, pipes and other construction material.	Building or road construction and repair.
Sewage treatment residue	Solids from screens, grit chambers, septic tanks and sludge. The waste is highly odorous.	Septic tank and treatment plants.

typical moisture content of MSW may vary from 15 to 40 percent depending upon the composition of the waste and the climatic conditions.

The density of MSW depends upon the composition and degree of compaction. The uncompacted density of MSW is around 150 kg/m³ (250 lb/yd³). The density of collected solid waste is 235–350 kg/m³ (400–600). The energy content of MSW as collected is 9,890 kJ/kg (4,260 BTU/lb). Information on chemical composition of the organic portion of MSW is important for many processes such as incineration, composting, biodegradability, leachate generation, and others. The ultimate analysis of the organic fraction of MSW is in terms of the constituents carbon, hydrogen, oxygen, nitrogen, sulfur and ash. These values are summarized in Table 2.6. The values, although considered typical, may have wide variations depending upon the moisture content and other components.

2.5.2 COLLECTION AND TRANSPORT

Solid waste collection and transport involves storage at the generation and pick-up points, pick up by the crew, trucks driving around the

TABLE 2.4. Distribution of Municipal Solid Wastes.

Category	Source	Percent by Weight
Residential	Single and multiple family residential areas including apartments. Wastes generated are food wastes, paper, cans, bottles, packaging, yard wastes.	40–55
Commercial	Stores, small industries* (non-hazardous wastes) restaurants, markets, office buildings, hotels, shopping areas, and other establishments. Wastes generated are food wastes, rubbish, ash, wood scraps.	15–30
Institutional	Schools, hospitals, city hall, community centers, prisons. The waste is similar to that of commercial sources.	2–5
Municipal services	Parks, streets, beaches, other recreational areas. The waste is generated from open areas, sidewalk and street cleaning, and landscaping. The waste contains both combustibles and non-combustibles such as leaves, paper, litter, grass, branches, dirt, ash, catch-basin dirt, etc.	10–15
Construction and demolition	Debris and large discards. The contents may include wood, metal pieces, boxes, dirt and debris.	5–10
Treatment plant residues*	Water and wastewater treatment plant sludge cakes, screenings, grit, scum.	0–5
	Total	4.0

*Some industrial and treatment plant residues are not collected with municipal solid wastes.
Source: Adapted in parts from Tchobanoglous et al. (1993), U.S. EPA (1988, and 1990).

TABLE 2.5. Typical Composition of MSW and
Moisture Content of Each Component.

Component	Percent by Weight		Percent Moisture Content	
	Range	Typical	Range	Typical
Food wastes	4–10	7	50–80	70
Paper and paperboard	18–60	40	4–10	6
Plastics	4–10	8	1–4	2
Textiles	0–4	2	6–15	10
Rubber	0–2	1	1–4	2
Leather	0–2	2	8–12	10
Yard wastes	0–20	18	30–80	60
Wood	1–6	3	15–40	20
Organics	0–5	2	10–60	25
Glass	4–16	7	1–4	2
Aluminum	1–3	2	2–4	3
Other metals	0–1	1	2–4	2
Ferrous metals	2–10	6	2–6	3
Miscellaneous inorganics	0–5	2	6–12	8
Municipal solid waste		100	15–40	20

Source: Adapted in part from U.S. EPA (1988, 1989, and 1990), Tchobanoglous et al. (1993), Peavy et al. (1985), and Matrecon, Inc. (1980).

TABLE 2.6. Typical Ultimate Analysis of Dry Combustible
Components of MSW.

Component	Percent Dry Weight Basis					
	Carbon	Hydrogen	Oxygen	Nitrogen	Sulfur	Ash
Food wastes	48.0	6.4	37.6	2.6	0.4	5.0
Paper and cardboard	44.0	6.0	44.0	0.3	0.2	5.5
Plastics	60.0	7.2	22.8	—	—	10.0
Textiles	55.0	6.6	31.2	4.6	0.2	2.5
Rubber	78.0	10.0	—	2.0	—	10.0
Leather	60.0	8.0	11.6	10.0	0.4	10.0
Garden trimmings	48.0	5.7	37.6	3.4	0.3	5.0
Wood	49.5	6.0	42.7	0.2	0.1	1.5
Total combustibles in MSW	46.0	6.1	41.0	0.9	0.2	5.8

Adapted in part from Tchobanoglous et al. (1993), Peavy et al. (1985), Matrecon, Inc. (1980).

18

neighborhood, and truck transport to a transfer station or disposal point. The collection is difficult, complex and costly. Collection of solid waste typically consumes 60–80 percent of the total solid waste budget of a community. Therefore, any improvement in the collection system can reduce overall cost significantly (U.S. Congress 1989).

2.5.2.1 Onsite Storage and Handling

In single family residential areas solid waste storage is handled by residents and tenants. Commonly used containers are plastic or galvanized metal containers, and disposable paper or plastic bags. The plastic or galvanized containers are 75–150 liter size with tight covering. The single use paper or plastic bags are generally used when curb service is provided and the homeowner is responsible for placing the bags along the curb. In high-rise buildings the waste is picked up by the building maintenance personnel, or special vertical chutes are provided to deliver the waste to a central location for storage, processing, or resource recovery. A recent development is to provide underground pneumatic transport systems to move waste to a central location for onsite storage, processing, or resource recovery. Apartment districts utilize stationary container systems into which the residents drop the solid wastes.

Solid wastes from commercial buildings are collected in large containers that may be stationary or transportable. Figure 2.4 shows various types of storage systems used for MSW.

2.5.2.2 Collection of Solid Wastes

In residential areas, the most common collection methods are curb or alley, setout, setout-setback, and backyard carry. In curb or alley service, the residents carry the single-use plastic bags and containers to the curb or collection point, and then return the empty container after pickup. Setout service utilizes a crew that carries the containers to the collection point. A separate collection crew empties the containers and residents return the empty containers. In setout-setback service, a third crew returns the empty containers. In backyard carry service, the collection crew transfers the solid waste into a wheeled barrel, then unloads it into the collection truck. The containers remain in the backyard.

Many communities have instituted regulations for separation of solid wastes at the source by the residents. Components such as newspapers and cardboard, aluminum, mixed glass, and food wastes from restaurants have been separated at the source. Although the concept is good, the participation of the public drops quickly. Also, the price of recycled material fluctuates greatly, and it is often more expensive to recycle waste material. All

(i)

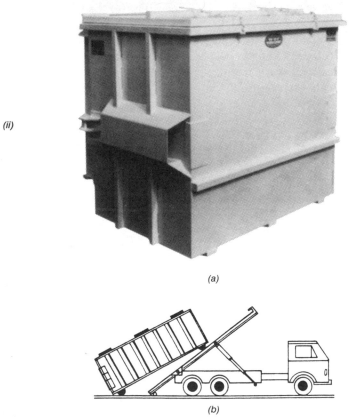

(ii)

(a)

(b)

Figure 2.4 Solid waste containers (a) stationary, (i) residential, (ii) commercial (courtesy Western Systems and Fabrication), and (b) transportable roll-off container (U.S. EPA 1976).

these factors are considered for instituting a mandatory separation and recycling program. There is, however, much interest in recycling these days due to mounting pressures of decreasing landfill sites, environmental concerns, economic incentives, and political support. Component separation methods of solid waste and recycling practices are presented further in Section 2.4.3.

The usual vehicle for residential collection of solid wastes is the manually rear or side-loaded compaction truck operating with a crew of two or three, including the driver. The typical truck is 14 to 18 m³ (15 to 20 yd³), and can carry 4 to 5 tons of wastes to the disposal site or transfer station. Woods (1992) reported features of new trucks that are equipped with an electronic control system for efficient operation and information storage and retrieval. Large self-loading compactor vehicles are equipped to unload several stationary containers serving apartments and shopping centers, and then replace the empty ones for reuse at the site. Other container trucks provide container exchange service. They are equipped to carry an empty storage container to a collection point, pick up a full container and transport it to a central location or disposal site, then replace the empty container at a new location. Various types of collection trucks are shown in Figure 2.5.

The frequency of solid waste collection in most communities is once or twice per week. The daily truck routes are fixed and balanced to provide a fair day's work. Several methods are used to optimize the route. Shuster and Schur (1974) have developed heuristic routing rules.

Liebman et al. (1975) presented a refuse collection model for routing vehicles in residential areas. A computerized Collection Management Information System (COLMIS) was developed to enable managers to evaluate, plan, and operate an effective and efficient solid waste collection system (U.S. EPA 1974).

2.5.2.3 Transfer Station and Transport

If the disposal site is too far from the city, the time spent by the crew of the pickup truck in unproductive travel becomes excessive. As a result, it may be uneconomical to use collection trucks for travel to the disposal site. Transfer stations are therefore established at convenient locations, and one-person trailer or large trucks, 27 to 46 m³ (35 to 60 yd³) or larger, are used to transport wastes to the disposal site. Henry and Heinke (1989) reported that long-haul trailer units are more economical if average round-trip haul distance is more than 50 km (30 miles). Among the important considerations in planning and designing a transfer station are location, type of station, access, and environmental effects.

At the transfer station, partial or complete solid waste processing such

(i)

(ii)

(a)

Figure 2.5 Various types of collection trucks used for collection and transport of solid waste to a central location or disposal site (a) trucks for collection of residential solid wastes, (i) rear loaded (courtesy The Heil Co.), (ii) automatic side loaded (courtesy The Heil Co.).

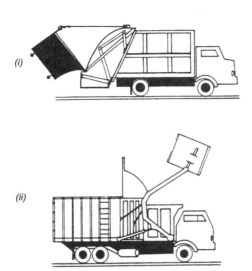

(i)

(ii)

(b)

(c)

Figure 2.5 (continued) (b) Truck to unload large container at the site (U.S. EPA 1976) (i) rear loading (ii) front loading, (c) tilt-frame vehicle to transport containers (courtesy Huge Haul, Inc.).

as sorting, shredding, compacting, baling, or composting may be provided. The objective is to reduce the volume, alter the physical form, and recover usable materials.

It is important that the transfer station be located as near as possible to the generation center. Good access roads as well as secondary or supplemental means of transportation are necessary. Also, the site must be environmentally acceptable. If more than one transfer station and disposal sites are used, then optimum allocation of wastes from each transfer station to each disposal site will be necessary. This is a classical problem in the field of operations research. Several mathematical models can be used to achieve economical solid waste allocation to the transfer stations and distribution to the disposal sites. Tchobanoglous et al. (1993), Davis and Cornwell (1991), and Vesilind et al. (1988) provided examples to optimize solid waste allocation and distribution.

2.5.3 RESOURCE RECOVERY AND RECYCLING

Many components of municipal solid wastes can be reused as secondary material. Among these are paper, cardboard, plastic, glass, ferrous metal, aluminum, and other nonferrous metals. These materials must be separated from MSW before they can be recycled. In this section, material recycling, and separation methods are first briefly presented, followed by bioconversion and refuse derived fuel (RDF) methods.

2.5.3.1 Material Recycling

The thrust of much state and local legislative action in the mid-1980s was to encourage recycling of various components of municipal solid waste. Many efforts have been made in the area of recycling, but have not been very successful. There are several reasons:

(1) Virgin raw materials, with which recycled material must compete for market share, are of known quality, and have little or no contamination.
(2) Recycled material may contain increased levels of foreign materials that could interfere with product quality.
(3) Uncertainty of supply and price variation of secondary material
(4) Methods of quality control of recycled material are not developed as for virgin materials.

Many components of MSW are currently recycled. Among these are paper and paper products. These products are recycled in manufacturing building materials such as roofing felt, insulation and wallboard, and are also used to manufacture cartons and containers. Plastic is recycled to produce insulating material, sheets, bags, and structural material. Energy is recovered from combustion of organic wastes. Other components of MSW being recycled are aluminum, glass, and ferrous metal. The U.S. EPA (1990) estimates that approximately 13 percent of MSW is currently recycled, and a 14 percent fraction is incinerated. Gross discards and percent recovery of various components of municipal solid wastes generated in 1988 are provided in Table 2.7.

2.5.3.2 Separation Methods

The separation of material is performed by the users at the source, or separated from mixed refuse at a central processing facility. Material sepa-

TABLE 2.7. Generation, Materials Recovery and Composting, and Discards of Materials in Municipal Solid Waste, 1988 (Millions of Tons and Percent).

Materials	Generation		Recovery for Recycling & Composting		Discards*	
	Million Tons	% of Total Generation	Million Tons	% of Material Generation	Million Tons	% of Total Discards
Paper and Paperboard	71.8	40.0	18.4	25.6	53.4	34.2
Glass	12.5	7.0	1.5	12.0	11.0	7.1
Metals						
Ferrous Metals	11.7	6.5	0.7	5.8	10.9	7.1
Aluminum	2.5	1.4	0.8	31.7	1.7	1.1
Other Nonferrous Metals	1.1	0.6	0.7	65.1	0.4	0.3
Total Metals	15.3	8.5	2.2	14.6	13.1	8.4
Plastics	14.4	8.0	0.2	1.1	14.3	9.1
Rubber and Leather	4.6	2.5	0.1	2.3	4.4	2.9
Textiles	3.9	2.1	Neg.	0.6	3.8	2.5
Wood	6.5	3.6	Neg.	Neg.	6.5	4.2
Other	3.1	1.7	0.7	21.7	2.4	1.6
Total Nonfood Product Wastes	132.1	73.5	23.1	17.5	109.0	69.9
Other Wastes						
Food Wastes	13.2	7.4	Neg.	Neg.	13.2	8.5
Yard Wastes	31.6	17.6	0.5	1.6	31.1	20.0
Misc. Inorganic Wastes	2.7	1.5	Neg.	Neg.	2.7	1.7
Total Wastes	179.6	100.0	23.5	13.1	156.0	100.0

*Discards after recovery for recycling and composting, but before combustion.
Neg. = Negligible.
Details may not add to totals due to rounding.
Source: U.S. EPA (1990).

RECYCLING

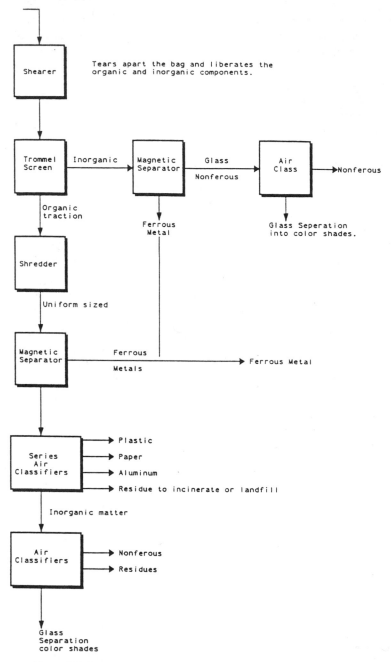

Figure 2.6 Generalized process train for material separation and recovery.

ration at the source involves users separating the material into different components, followed by transporting from the point of generation to a secondary material dealer. Unfortunately, active household response for separation at the source has been very poor. As a result, effort has been directed toward separation of MSW at a central facility.

Handpicking is a long-used form of separation of a few components of solid wastes. In this operation, a conveyor moves the solid waste past by a group of workers who pick up the designated components by hand. This method of separation is costly, and only a few bulky components, such as bundled newspapers and cardboard, can be separated.

A mechanized material recovery method provided by Wilson (1981) utilizes shearers that break open the bags and liberate cans and bottles. Trommel screens separate cans, glass and other inorganic material. The organic material is shredded and passed through air classifiers, which separate the components desired for recovery of fibers for paper making or for producing refuse derived fuel. Magnetic and electromechanical systems separate ferrous and nonferrous metals. Glass is sorted in different shades of color for recycling. Figure 2.6 is a generalized process train for material separation and recovery. Drobny et al. (1971), Wilson (1981), Vesilind and Rimer (1981), Peavy et al. (1985), Davis and Cornwell (1991), and Tchobanoglous et al. (1993), provided theory and design information on shearers, trommel screens, shredders, magnetic separators, air classifiers, Sortex color glass separator, and Franklin paper fiber recovery system. Many of these material separation devices are shown in Figure 2.7.

2.5.3.3 Bioconversion

Bioconversion of the organic fraction of municipal solid waste into a number of products including sugar, ethanol, protein compost and methane, has been reported in the literature (Drobny et al. (1971); U.S. EPA (1976); Wilson (1981); Vesilind and Rimer (1981); and Tchobanoglous et al. 1993).

2.5.3.3.1 Sugar

The recovery of fibers from paper has cellulose as the major constituent. The cellulose is hydrolyzed into sugars. The hydrolysis of cellulose produces glucose and mixtures of other sugars. Hydrolysis of paper fibers is achieved under low pH, or by enzymes. The hydrolysis reaction is given by Equation (2.1) (see page 31).

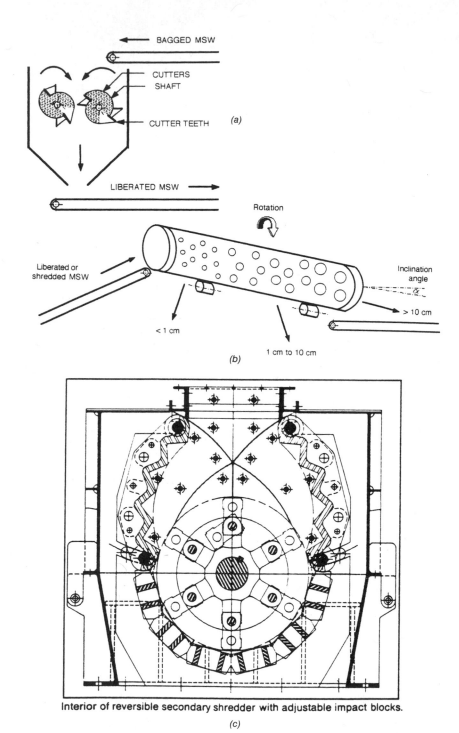

Interior of reversible secondary shredder with adjustable impact blocks.

(c)

Figure 2.7 Material separation equipment (a) rotary shearer, (b) trommel screen, (c) shredder (hammermill) (courtesy Williams Patent Crusher and Pulverizer Co.).

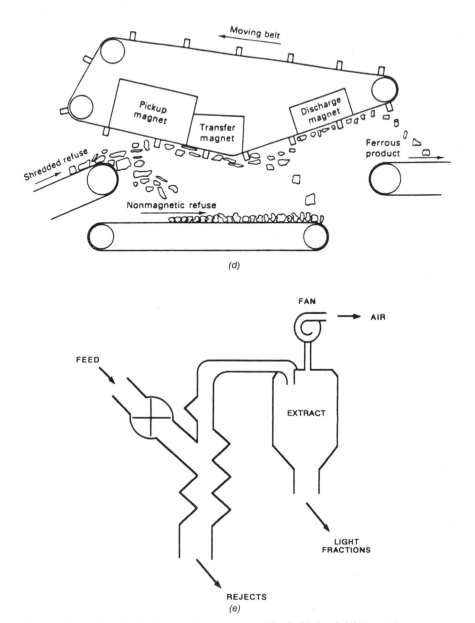

(d)

(e)

Figure 2.7 (continued) (d) Magnetic belt separator, John T. Pfeffer, *Solid Waste Management Engineering*, © 1992, p. 152. Reprinted by permission of *Prentice-Hall*, Englewood Cliffs, New Jersey, (e) air classifier.

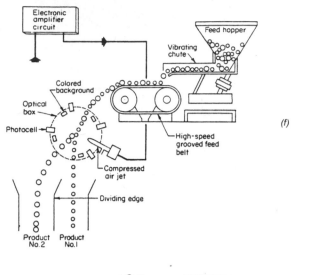

(f)

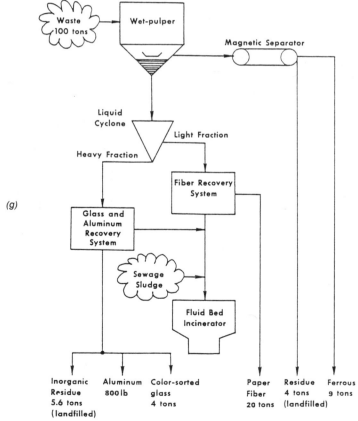

(g)

Figure 2.7 (continued) (f) *Sortex* optical separator (Drobny et al., 1971) (g) *Black Clawson* hydropulper for fiber recovery (U.S. EPA 1976).

$$(C_6H_{10}O_5)_n + nH_2O \xrightarrow[\substack{3447 \text{ kN/m}^2 \text{ (500 psi)} \\ 150-180°C \\ \text{or} \\ \text{Enzyme}}]{H_2SO_4} nC_6H_{12}O_6 \qquad (2.1)$$

2.5.3.3.2 Ethanol and Other Fermentation Products

Microorganisms can produce alcohols and short-chain organic acids. These products have commercial value in the industrial chemical market. Some of these fermentation reactions are given by Equations (2.2)–(2.5).

$$C_6H_{12}O_6 + 2H_2O \longrightarrow 2CH_3CH_2OH + 2H_2CO_3 \qquad (2.2)$$
$$\text{Ethanol}$$

$$2CH_3CH_2OH + 2H_2O \longrightarrow 2CH_3COOH + 4H_2 \qquad (2.3)$$
$$\text{Acetic acid}$$

$$CH_3COOH + H_2O \longrightarrow CH_4 + H_2CO_3 \qquad (2.4)$$
$$\text{Methane}$$

$$4H_2 + H_2CO_3 \longrightarrow CH_4 + 3H_2O \qquad (2.5)$$
$$\text{Methane}$$

Many researchers have investigated the production of commercial quantities of these products using paper fibers recovered from municipal solid wastes. However, the cost of production and recovery of these products in a relatively pure state from MSW is much greater than the cost of producing the raw materials.

2.5.3.3.3 Protein

Earlier research conducted by the Denver Research Institute suggested that fibers produced from waste paper can be converted to protein using a fast-growing-hydrocarbon-cellulose digesting organism. The organism is *Candia utilis* culture, commonly called *Torula yeast*. The protein thus produced is usable by livestock (Drobny et al., 1971).

Another concept, reported by Wilson (1981), is to utilize processed organic wastes as the feedstock for a rapidly expanding population of red earthworms. This process is called *annelidic recycling*. The organic component is converted into a rich fertilizer in the form of worm castings, the excess worms are dried into high-protein food supplement in animal food. Another similar concept is to breed flies and recover larvae as a protein source for animal feed supplement.

2.5.3.3.4 Composting

Composting involves conversion of the organic component of solid waste into humus-like material to be used as soil conditioner. Both aerobic and anaerobic microbial decomposition may be used. The composting process involves (1) preparation of solid waste, (2) decomposition, and (3) product preparation for marketing. The solid waste preparation involves receiving, sorting, separating, shredding; removing of plastic, rubber, leather and the like; and moisture and nutrient addition. The decomposition is achieved by microorganisms, mostly under aerobic condition. The simplest method is windrow composting, in which long piles are prepared that are 1.5–2.0 m high, and 2.0–2.5 m wide. The piles are turned regularly to ensure adequate aeration. Finished compost is ready in two to three weeks. Many proprietary composting processes utilize mechanical digesters with forced aeration, seeding, moisture and nutrient adjustment to accelerate compost production to less than a week. Among these processes are (1) rotating drum digester, (2) vertical tower in which air is supplied from the bottom and mechanical agitation is provided by moving arm, and (3) silo digester resembling multiple-hearth incinerator consisting of a number of tiers (Wilson 1981). Peavy et al. (1985) provided a generalized equation for aerobic decomposition of organic matter during the composting process [Equation (2.6)].

$$C_aH_bO_cN_d + 0.5 \ (ny + 2s + r - c)O_2 \quad \longrightarrow$$

$$n(C_wH_xO_yN_z) + sCO_2 + rH_2O + (d - nz)NH_3 \tag{2.6}$$

Where $C_aH_bO_cN_d$ and $C_wH_nO_yN_z$ represent the generalized chemical formula for representing organic matter before and after decomposition.

$$r = 0.5 \ [b - nx - 3 \ (d - nz)]$$
$$s = a - nw$$

The finished compost is a dark brown or black color material, has low carbon-nitrogen ratio, and has high capacity for base exchange and for water absorption. The finished product is obtained by fine grinding, blending with nutrients and various additives, and bagging and storage.

2.5.3.4 Anaerobic Digestion

Anaerobic digestion is principally used for methane generation. There are three basic steps: (1) preparation of organic fraction, (2) digestion, and (3) gas recovery and disposal of digested residues. The solid waste preparation step is similar to that in composting. Digestion is achieved in enclosed continuous flow mixed reactors that are heated to about 55 to 60°C. The digestion period may vary from 5 to 10 days. The recovery of methane involves capture, purification, storage, and piping. Tchobanoglous et al. (1977) and Peavy et al. (1985) suggested a generalized equation for anaerobic breakdown of organic matter in the solid waste.

$$C_aH_bO_cN_d \longrightarrow$$

$$n(C_wH_xO_yN_z) + mCH_4 + sCO_2 + rH_2O + (d - nz)NH_3$$

(2.7)

Where

$$r = c - ny - 2s$$
$$s = a - nw - m$$

Other terms are defined in Equation (2.6).

The digested residue may be dewatered and converted into compost, or applied over farmland without dewatering. Recovery of gas from municipal landfill is technically feasible and economically justifiable (Ham et al. 1979; and Robinson 1986). Gas recovery systems from sanitary landfills are described in greater detail in Section 3.10.2.

2.5.3.5 Incineration and Energy Recovery

Incineration of MSW is practiced to reduce waste volume and recover energy. The batch-fed incinerators built in the 1930s and 1940s reduced the volume but were major contributors to air pollution problems. Most of these incinerators have been either shut down or replaced by newer designs.

The newer incinerators utilize innovative technology to produce steam more efficiently and reduce air pollutants to a greater extent. The capital and operating costs, however, are quite high. Henry and Heinke (1989) reported capital cost of about $120 million per 1000 tons of daily capacity, and operating cost of $15 to $30 per ton, for cities over 300,000 population (1987 dollars). The unit cost of incinerators for smaller cities is even larger. The high cost of installing the air pollution control equipment forced municipalities to seek cheaper methods such as sanitary landfills for solid waste disposal. Due to stricter regulations on landfilling, and rapidly diminishing capacity, there is a renewed interest in incinerator design and construction. Davis and Cornwell (1991) reported that in 1978 there were only 56 facilities that remained active in the United States. Today there are more than 100 facilities in operation, and over 100 more are in the planning and construction stages.

2.5.3.5.1 Incineration Process

Incineration involves combustion of the organic fraction of the solid waste. The basic requirements of incineration are:

- Air and fuel must be mixed in proper proportion.
- Air and the combustible gases, and fuel (if needed), must be mixed adequately.
- Temperature must be sufficiently high for ignition and combustion of both the solid waste and the gas component.
- Furnace volume must be large enough to provide the retention time needed for complete combustion.
- Furnace proportion must be such that ignition temperatures are maintained and fly ash entrainment is minimized.

The combustion temperature must be higher than 800°C for odor-free burning. Also, at temperatures higher than 1200°C the fusion of ash will occur. Conventional refractory wall incinerators utilize air in excess of theoretical air requirement for complete oxidation of organic matter. Air in excess of 300 percent of the theoretical air has been used in earlier incinerators. Combustion of organic matter at higher excess air produces larger quantities of combustion products, and burning temperatures are lower. Newer incinerators utilize less excess air, or may even be operated in an oxygen-starved condition for gasification (partial combustion to produce combustible fuel).

Incinerators are single chamber or multiple chamber. In a single-chamber incinerator, solid waste drying, air mixing (under fire or over fire), ignition, and combustion takes place in a single chamber. Several types of continuous-feed stokers are used to feed the solid waste into the burning area or grate. In multiple chamber incinerators, the volatile components of the solid wastes are vaporized and partially oxidized while passing from an ignition chamber through the flame port into the connected mixing chamber. In the mixing chamber, the volatile components of solid wastes and products of combustion are mixed with secondary air. Turbulent mixing is achieved from abrupt changes in flow direction accompanied by flow through restricted flow areas. In the combustion chamber, the gases undergo additional changes in direction accompanied by expansion and final oxidation of the combustible components. Finally, the fly ash and solids are removed by wall impingement and simple settling.

2.5.3.5.2 Energy Recovery

There are several different technologies that are presently available to recover energy from solid wastes. In conventional refractory lined incinerators, waste heat boilers are installed to extract heat from the combustion gases at lower excess air. Hot gases from the burning refuse provide energy to generate steam that can be passed through a standard turbine generator for electricity production. The gases are cooled, waste heat is recovered for steam production, and the volume of the off gases is reduced due to cooling before reaching the air pollution control equipment. The flow diagram is shown in Figure 2.8 (Knight 1988).

In water-wall incinerators, the walls of the combustion chamber are lined with boiler tubes. The circulating water absorbs heat generated in the combustion chamber. Thus, even in a refractory lined combustion chamber, the temperature can be controlled without introducing excess air. The need for air pollution control equipment is significantly reduced. Corrosion of water tubes is generally a problem due to hydrochloric acid produced from the burning of some plastics. Nasenbeny (1987) provided a list of the nation's largest 10 resource recovery plants (Table 2.8).

2.5.3.5.3 Pyrolysis

Pyrolysis is thermal cracking or destructive distillation of organic matter. In this process, the organic matter is heated in an oxygen-free or

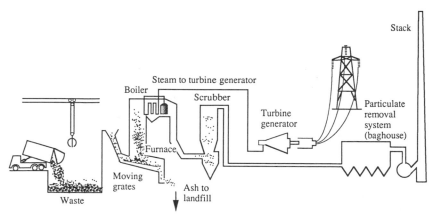

Figure 2.8 Incineration of municipal solid wastes with power generation from off gases. Source: Adapted from Knight (1988).

TABLE 2.8. Nation's Largest Resource Recovery Plants.

Location	Capacity (tons per day)
Detroit, MI	4,000
Dade County, FL	3,000
Peekskill, NY	2,250
Baltimore, MD	2,250
Hartford, CT	2,000
Honolulu, HI	2,000
Pinellas County, FL	2,000
Saugus, MA	1,500
Niagara Falls, NY	1,500
Lawrence, MA	1,300
Chicago, IL	1,250

Source: Adapted from Nasenbeny (1987) with permission from *Consulting Specifying Engineer*.

oxygen-starved atmosphere. Through thermal cracking and condensation, organic materials break down into three components:

(1) A gas consisting primarily of hydrogen, methane, and carbon monoxide

(2) A liquid fuel that includes organic chemicals such as acetic acid, acetone, and methanol

(3) A char consisting of almost pure carbon, plus many residues such as glass, metal, and inert material that may be present.

Equation (2.8) gives the generalized pyrolysis reaction for cellulose.

$$3(C_6H_{10}O_5) \longrightarrow$$

$$8H_2O + C_6H_8O + 2CO + 2CO_2 + CH_4 + H_2 + 7C$$

$$(2.8)$$

The liquid fraction is represented by C_6H_8O. The distribution of gaseous, liquid, and solid fraction varies greatly with the temperature at which pyrolysis is carried out (Tchobanoglous et al. 1993).

2.5.3.5.4 Incineration-Pyrolysis

Pyrolysis incineration technology has shown great promise. Several systems have been developed that generate storable, transportable fuel from solid wastes. *Union Carbide* developed a *Purox* system that uses pure oxygen, and the product gas is a clean burning fuel. In early 1970s, the *Garrett Research and Development Company* developed a flash pyrolysis system that produced an oil-like liquid. The *Envirochem Landguard (Monsanto)* system also developed at that time used a rotary kiln, and produced steam and combustible residue in the form of char. One full-scale *Occidental Flash Pyrolysis System* for MSW was constructed in El Cajon, California. This system produced pyrolytic oils, gases and char (Tchobanoglous et al. 1993). Discussions on these, and many other systems may be found in several references (Wilson 1981; Vesilind and Rimer 1981; Parkinson 1984; Nasenbeny 1987; and Knight 1988; Tchobanoglous et al. 1993).

2.5.3.5.5 Refuse-Derived Fuels (RDF)

Solid wastes may be burned directly in incinerators (a process called mass-burning) or converted to more efficient refuse-derived fuel (RDF). The solid waste is processed by means of size reduction and material separation techniques to obtain a product which has a substantial heat value. Physically, shredded and air classified organic fraction of solid wastes are burned along with a fossil fuel such as coal. Many boilers need only minor modifications to accept such fuel. Also, the shredded and air classified solid wastes can be densified into cubes or pellets that are suitable for many thermal conversion processes such as incineration, gasification, and pyrolysis. The process flow diagram of an RDF system is shown in Figure 2.9 (Knight 1988).

Many communities have built refuse-to-energy plants that are fast replacing sanitary landfills. The steam is used for heating and for driving turbines to produce electricity. Kiser (1992) reported that there are 190 municipal waste combuster plants in the USA. Of these, 142 facilities are

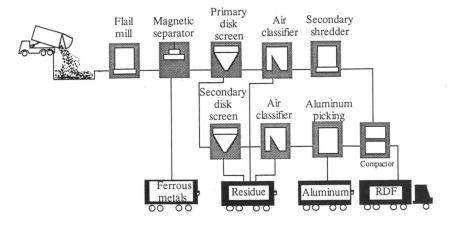

Figure 2.9 Refuse-derived fuel (RDF) system. Source: Adapted from Knight (1988).

waste-to-energy combusters. The total processing capacity exceeds 31.4 million tons of MSW per year. These facilities do not include RDF-processing plants or simple incinerators. There are also approximately 49 waste-to-energy combusters planned or under construction.

2.5.4 DISPOSAL BY LANDFILLING

The volume of municipal solid waste is greatly reduced by incineration, conversion processes or resource recovery. In all cases, there is a residue that must be disposed of so that it no longer creates a nuisance or hazard to the society. Engineering principles must be followed for site selection and design of ultimate-land disposal methods. An acceptable land disposal method of MSW and residues utilizes sanitary landfills.

Sanitary landfilling is the most common method of solid waste disposal in the United States. It is economical, and applies to all components of MSW. Proper site selection is perhaps the most difficult part of disposal by landfilling. The operation involves compaction of solid wastes in layers, then covering with a layer of compacted soil at the end of each day's operation. In recent years, special care has been required in site selection, refuse compaction, cover application, leachate collection and treatment, and site monitoring. The rest of this book is devoted to sanitary landfill design, construction and operation, and in particular, leachate generation, characterization, containment and treatment.

2.6 REFERENCES

Davis, M. L. and D. A. Cornwell. 1991. *Introduction to Environmental Engineering, Second Edition*, New York: MacGraw-Hill, Inc.

Drobny, N. L., H. E. Hull and R. F. Testin. 1971. *Recovery and Utilization of Municipal Solid Wastes*. EPA-SW-10C-71, U.S. Environmental Protection Agency, Office of Solid Waste Management Program, Washington, DC: (NTIS No. PB-204 922).

Ham, R. K., K. Hekimian, S. Katten, W. J. Lockman, R. J. Lofty, D. E. McFaddin and E. J. Daley. 1979. *Recovery, Processing, and Utilization of Gas from Sanitary Landfills*. EPA-600/2-79-001, U.S. Environmental Protection Agency, Cincinnati, OH, p. 133.

Henry, G. and G. W. Heinke. 1989. *Environmental Science and Engineering*. Englewood Cliffs, NJ: Prentice Hall.

Kiser, J. V. L. 1992. "Municipal Waste Combustion in North America: 1992 Update," *Waste Age*, 23(11):26–34.

Knight, A. 1988. "Energy from Waste: Recovering a Throwaway Resource," *EPRI Journal*, October/November, pp. 25–35.

Liebman, J. C., J. W. Male and M. Wathne. 1975. "Minimum Cost in Residential Refuse Vehicle Route," *Journal of the Environmental Engineering Division, American Society of Civil Engineers*, 101(EE3):399–412.

Matrecon, Inc. 1980. *Lining of Waste Impoundment and Disposal Facilities,* SW-870, U.S. Environmental Protection Agency, Cincinnati, OH.

Nasenbeny, R. J. 1987. "Refuse-to-Energy Plant Gets City Out of Dumps," *Consulting Specifying Engineer,* March:54–56.

Parkinson, G. 1984. "Don't Discard Garbage as a Source of Energy," *Chemical Engineering,* March 19:38–45.

Peavy, H. S., D. R. Rowe and G. Tchoganoglous. 1985. *Environmental Engineering,* New York: McGraw-Hill, Inc.

Pfeffer, G. T. 1992. *Solid Waste Management Engineering.* Englewood Cliffs, NJ: Prentice Hall, Inc.

Robinson, W. D. 1986. *The Solid Waste Handbook, A Practical Guide,* New York: John Wiley and Sons.

Shuster, K. A. and D. A. Schur. 1974. *Heuristic Routing for Solid Waste Collection Vehicles,* SW-11, U.S. Environmental Protection Agency.

Tchobanoglous, G., H. Theisen and R. Eliassen. 1977. *Solid Wastes, Engineering Principles and Management Issues.* New York, NY: McGraw-Hill Book Co.

Tchobanoglous, G., H. Theisen and S. Vigil. 1993. *Integrated Solid Waste Management, Engineering Principles and Management Issues.* New York: McGraw Hill, Inc.

U.S. Congress. 1965. The Solid Waste Disposal Act, Title II of Public Law 89-272.

U.S. Congress. 1989. *Facing America's Trash: What Next for Municipal Solid Waste.* OTA-0-424, Office of Technology Assessment, Washington, DC: U.S. Government Printing Office.

U.S. Environmental Protection Agency. 1974. *User's Manual for COLMIS: A Collection Management Information System for Solid Waste Management, Vol. 1,* SW-57C, Washington, DC: Office of Solid Waste Management Program.

U.S. Environmental Protection Agency. 1976. *Decision-Makers Guide in Solid Waste Management.* SW-500, Washington, DC: U.S. Government Printing Office.

U.S. Environmental Protection Agency. 1988. *Characterization of Municipal Solid Waste in the United States, 1960 to 2000,* Update 1988, Final Report Prepared by Franklin Associates, Inc.

U.S. Environmental Protection Agency. 1989. *Decision-Makers Guide in Solid Waste Management,* EPA/530-SW 89-072, Washington, DC.

U.S. Environmental Protection Agency. 1990. *Characterization of Municipal Solid Waste in the United States: 1990 Update,* EPA/530-SW-90-042, Washington, DC: Solid Waste and Emergency Response.

U.S. Environmental Protection Agency. 1991. *Solid Waste Disposal Facility Criteria; Final Rule,* 40 CFR Part II, 257 and 258, Subtitle D of the Resource Conservation and Recovery Act (RCRA), 258.40, Design Criteria.

U.S. Environmental Protection Agency. 1992. 40 CFR 260 and 261, July 1, 1992, pp. 4–131.

Vesilind, P. A., J. J. Pierce and R. F. Weiner. 1988. *Environmental Engineering, Second Edition.* Boston, MA: Butterworth's.

Vesilind, P. A. and A. E. Rimer. 1981. *Unit Operations in Response Recovery Engineering.* Englewood Cliffs, NJ: Prentice Hall, Inc.

Wentz, C. A. 1989. *Hazardous Waste Management.* New York: McGraw-Hill Book Co.

Wilson, D. C. 1981. *Waste Management: Planning, Evaluation Technologies.* Oxford: Clarendon Press.

Woods, R. 1992. "Refuse Vehicles of the '90s," *Waste Age,* 23(5):38–44.

CHAPTER 3

Sanitary Landfill Planning, Design and Operation

3.1 INTRODUCTION

SANITARY landfilling is the most common method of ultimate disposal of solid wastes. It is an engineered method of disposing solid wastes on land in a manner that minimizes environmental hazards and nuisances. At a site that is carefully selected, designed and prepared, the municipal solid waste is spread in thin layers, compacted to the smallest practical volume, and covered with compacted earth at the end of each day of operation.

The planning, design and operation of a modern sanitary landfill involves complex scientific, engineering, and economic principles. The design and operation of existing and new municipal solid waste landfills are strictly regulated under the *Subtitle D Rule* (U.S. EPA 1991a). The new Federal Standards for municipal landfills are an effort to control leachate from municipal landfills that could pollute the groundwater supplies and soils.

In this chapter, the basic factors for a sanitary landfill site selection, facility design, and operation are covered. Subsequent chapters in this book are devoted to solid waste and hazardous waste legislation on landfilling practices, leachate generation, migration, containment, and treatment of leachate from municipal solid waste landfills.

3.2 SITE SELECTION

The site selection of a sanitary landfill is perhaps the most difficult and controversial issue of the planning process. Davis and Cornwell (1991) listed 15 basic factors that should be considered during the site selection process. These are:

(1) Public opposition
(2) Proximity of major roadway

41

(3) Speed limits

(4) Load limits of roadway

(5) Bridge capacities

(6) Underpass limitations

(7) Traffic patterns and congestion

(8) Average haul distance or haul time

(9) Detours

(10) Hydrology

(11) Availability of cover material

(12) Climate (for example, floods, mud slides, snow)

(13) Zoning requirements

(14) Buffer areas around the site

(15) Historic buildings, endangered species, wetlands, and other environmental factors

Davis and Cornwell (1991) also listed other siting requirements such as site not less than (1) 30 m from streams, (2) 160 m from drinking water wells, (3) 65 m from houses, school and parks, and (4) 3,000 m from airport runways.

Ideally, a sanitary landfill site should be inexpensive land within an economical hauling distance, good and year-round access, soil of low permeability well above the groundwater table, and the area should be well drained. O'Leary, Tansel and Fero (1986), Pfeffer (1992), and Tchobanoglous et al. (1993) provided many important site specific charateristics of a sanitary landfill. Some of these features are summarized in Table 3.1.

TABLE 3.1. Sanitary Landfill Site Specific Characteristics.

Factor	Degree of Limitation		
	Severe	Moderate	Minimal
Land slope	>15%	3–15%	>3%
Surface deposits	Clean sand/gravel	Sand/gravel with silt	Silty clay
Bedrock depth	Clay >10 ft	10–25 ft	>25 ft
Bedrock type	Fractured Limestone	Sandstone	
Groundwater depth	<10 ft	10–20 ft	>25 ft
Distance to:			
Water well	<300 ft	300–1000 ft	>1000 ft
Floodplain	<300 ft	300–1000 ft	>1000 ft
Stream/lake	<1000 ft		>1000 ft
Parks	<1000 ft		>1000 ft
Wetlands	Located within		

m = ft × 0.3048.

Source: O'Leary, Tansel and Fero (1986). Adapted with permission from *Waste Age*.

3.3 SOIL AND GEOLOGY

Understanding geologic conditions of the area is essential. The type of soil available at the site will be an important factor in the design and operation of a sanitary landfill. Among these are chemical properties of bed rock, soil type particularly as it may relate to the movement of water and gas, permeability, workability, and vegetation are essential elements of soil evaluation, as are stratigraphy and structure of the bed rock.

3.3.1 SOIL CLASSIFICATION AND PROPERTIES

There are two soil classification systems that are commonly used (Brunner and Keller 1972). These are U.S. Department of Agriculture Soil Conservation Service (USDA), and The Unified Soil Classification System (USCS). These classifications are provided in Figure 3.1 and Table 3.2. Information on soil classification systems for landfill application may be obtained from several sources (Meyer and Knight 1961; Brunner and Keller 1972; Noble 1976; Lutton et al. 1979).

Clay soils are very fine in texture even though they may contain small to moderate amounts of silt and sand. Their physical and chemical properties may vary greatly. When dry, a clay soil can be hard and tough like a rock, and can support heavy loads. When wet, it often becomes very soft, is sticky or slippery, and is very difficult to handle. The permeability of wet clay is very low. Also, when clays are dry, they shrink and crack. When wet, they absorb large amounts of water and swell. Impervious clays are extensively used to construct impermeable linings.

3.3.2 SOIL COVER

The compacted soil cover is important in controlling the movement of leachate and surface water infiltration. The cover material is expected to serve many other functions such as controlling rodents, flies and scavenging birds, supporting vegetation, and restricting moisture and gas movement. Many of these functions and relative ratings of various soil types are presented in Table 3.3. The soil cover also serves as a road bed for vehicles. If water is to be kept out of the landfill, the soil must have low permeability. If landfill gases are to be vented, a reverse situation occurs. For road bed, the soil cover must be well drained. Therefore, soil cover material must be carefully evaluated based on the functions it must serve. There may be trade-offs between many desired functions. Generally peat and highly organic soils should be avoided. They are difficult to compact, are normally very sticky, and can vary in their moisture content.

TABLE 3.2. (left half) Unified Soil Classification System and Characteristics Pertinent to Sanitary Landfills.

Major Divisions		Symbol Letter	Hatching	Color	Name	Potential Frost Action	Drainage Characteristics*
COARSE-GRAINED SOILS	GRAVEL AND GRAVELLY SOILS	GW		RED	Well-graded gravels or gravel-sand mixtures, little or no fines	None to very slight	Excellent
		GP		RED	Poorly graded gravels or gravel-sand mixtures, little or no fines	None to very slight	Excellent
		GM		YELLOW	Silty gravels, gravel-sand-silt mixtures	Slight to medium	Fair to poor / Poor to practically impervious
		GC		YELLOW	Clayey gravels, gravel-sand-clay mixtures	Slight to medium	Poor to practically impervious
	SAND AND SANDY SOILS	SW		RED	Well-graded sands or gravelly sands, little or no fines	None to very slight	Excellent
		SP		RED	Poorly graded sands or gravelly sands, little or no fines	None to very slight	Excellent
		SM		YELLOW	Silty sands, sand-silt mixtures	Slight to high	Fair to poor / Poor to practically impervious
		SC		YELLOW	Clayey sands, sand-clay mixtures	Slight to high	Poor to practically impervious
FINE-GRAINED SOILS	SILTS AND CLAYS LL IS LESS THAN 50	ML		GREEN	Inorganic silts and very fine sands rock flour, silty or clayey fine sands or clayey silts with slight plasticity	Medium to very high	Fair to poor
		CL		GREEN	Inorganic clays of low to medium plasticity, gravelly clays, sandy clays, silty clays, lean clays	Medium to high	Practically impervious
		OL		GREEN	Organic silts and organic silt-clays of low plasticity	Medium to high	Poor
	SILTS AND CLAYS LL IS GREATER THAN 50	MH		BLUE	Inorganic silts, micaceous or diatomaceous fine sandy or silty soils, elastic silts	Medium to very high	Fair to poor
		CH		BLUE	Inorganic clays of high plasticity, fat clays	Medium	Practically impervious
		OH		BLUE	Organic clays of medium to high plasticity, organic silts	Medium	Practically impervious
HIGHLY ORGANIC SOILS		Pt		Orange	Peat and other highly organic soils	NOT RECOMMENDED FOR SANITARY LANDFILL CONSTRUCTION	

*Values are for guidance only; design should be based on test results.
†The equipment listed will usually produce the desired densities after a reasonable number of passes when moisture conditions and thickness of lift are properly controlled.
‡Compacted soil at optimum moisture content for Standard AASHO (Standard Proctor) compactive effort.
Source: Brunner and Keller (1972).

3.4 SITE PREPARATION AND DEVELOPMENT

The selected sanitary landfill site should have utilities such as electricity, telephone, potable water, water for fire fighting and dust control, sanitary sewer, all weather access road, fence, weighing station; and a building for

TABLE 3.2. (right half).

Value for Embankments	Permeability cm per sec	Compaction Characteristics †	Std AASHO Max Unit Dry Weight lb per cu ft †	Requirements for Seepage Control
Very stable, pervious shells of dikes and dams	$k > 10^{-2}$	Good, tractor, rubber-tired steel-wheeled roller	125-135	Positive cutoff
Reasonably stable, pervious shells of dikes and dams	$k > 10^{-2}$	Good,-tractor, rubber-tired steel-wheeled roller	115-125	Positive cutoff
Reasonably stable, not particularly suited to shells, but may be used for impervious cores or blankets	$k = 10^{-3}$ to 10^{-6}	Good, with close control, rubber-tired, sheepsfoot roller	120 -135	Toe trench to none
Fairly stable, may be used for impervious core	$k = 10^{-6}$ to 10^{-8}	Fair, rubber-tired, sheepsfoot roller	115-130	None
Very stable, pervious sections slope protection required	$k > 10^{-3}$	Good, tractor	110-130	Upstream blanket and toe drainage or wells
Reasonably stable, may be used in dike section with flat slopes	$k > 10^{-3}$	Good, tractor	100-120	Upstream blanket and toe drainage or wells
Fairly stable, not particularly suited to shells,but may be used for impervious cores or dikes	$k = 10^{-3}$ to 10^{-6}	Good, with close control, rubber-tired, sheepsfoot roller	110-125	Upstream blanket and toe drainage or wells
Fairly stable, use for impervious core for flood control structures	$k = 10^{-6}$ to 10^{-8}	Fair, sheepsfoot roller, rubber-tired	105-125	None
Poor stability, may be used for embankments with proper control	$k = 10^{-3}$ to 10^{-6}	Good to poor, close control essential, rubber-tired roller, sheepsfoot roller	95-120	Toe trench to none
Stable, impervious cores and blankets	$k = 10^{-6}$ to 10^{-8}	Fair to good, sheepsfoot roller, rubber-tired	95-120	None
Not suitable for embankments	$k = 10^{-4}$ to 10^{-6}	Fair to poor, sheepsfoot roller	80-100	None
Poor stability, core of hydraulic dam, not desirable in rolled fill construction	$k = 10^{-4}$ to 10^{-6}	Poor to very poor, sheepsfoot roller	70-95	None
Fair stability with flat slopes, thin cores, blankets and dike sections	$k = 10^{-6}$ to 10^{-8}	Fair to poor, sheepsfoot roller	75-105	None
Not suitable for embankments	$k = 10^{-6}$ to 10^{-8}	Poor to very poor, sheepsfoot roller	65-100	None
	NOT RECOMMENDED FOR SANITARY LANDFILL CONSTRUCTION			

storing hand tools and equipment parts, and for sanitary facilities. The site area and surface should be graded and prepared in accordance with the design plans and specifications. This may include excavation, filling, stockpiling of cover materials, construction of berms, and installation of leachate and gas control and monitoring systems.

3.5 EQUIPMENT

The equipment most commonly used in landfilling are crawler tractor, front-end loader, rubber-tired front-end loader, grader, scrapper, dragline

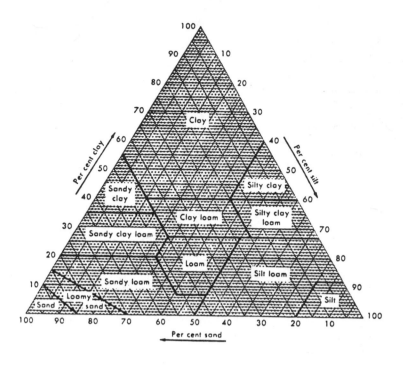

Sand—2.0 to 0.05 mm. diameter
Silt—0.05 to 0.002 mm. diameter
Clay—smaller than 0.002 mm. diameter

COMPARISON OF PARTICLE SIZE SCALES

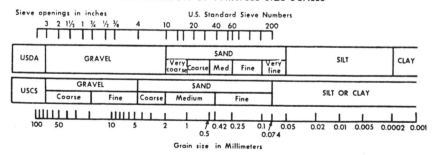

Figure 3.1 Textural classification chart (U.S. Department of Agriculture) and comparison of particle size scales [Source: Brunner and Keller (1972)].

TABLE 3.3. Suitability of General Soil Types as Cover Material.[a]

Function	Clean Gravel	Clayey Silty Gravel	Clean Sand	Clayey Silty Sand	Silt	Clay
Prevent rodents from burrowing or tunneling	G	F–G	G	P	P	P
Keep flies from emerging	P	F	P	P	G	E[b]
Minimize moisture entering fill	P	F–G	P	G–E	G–E	E[b]
Minimize landfill gas venting through cover	P	F–G	P	G–E	G–E	E[b]
Provide pleasing appearance and control blowing paper	E	E	E	E	E	E
Grow vegetation	E	G	P–F	E	G–E	F–G
Be permeable for venting decomposition gas[c]	E	P	G	P	P	P

[a]E, excellent; G, good; F, fair; P, poor.
[b]Except when cracks extend through the entire cover.
[c]Only if well drained.
Source: Brunner and Keller (1972).

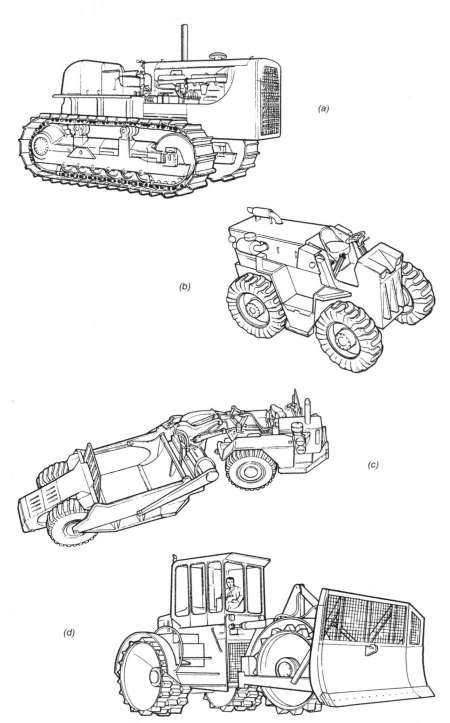

Figure 3.2 Common sanitary landfill equipment. (a) crawler tractor, (b) rubber-tired tractor, (c) scraper, (d) steel-wheel compactor.

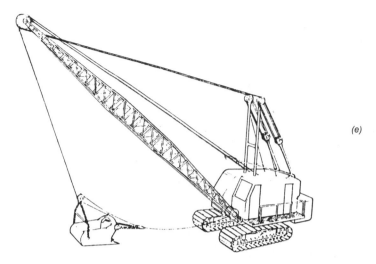

(e)

Figure 3.2 (continued) (e) Dragline [Source: Sorg and Hickman (1970)].

and compactor. This equipment is needed for landfilling and supporting services. The number and size of the equipment is based on the site condition and amount of solid wastes handled. Figure 3.2 shows some of the sanitary landfill equipment. The performance characteristics of sanitary landfill equipment and selection guide are provided in Tables 3.4 and 3.5.

3.6 METHODS OF LANDFILLING AND OPERATION

Sanitary landfilling involves unloading of solid waste, spreading, compacting, then covering it with at least 15 cm (6 inches) of compacted daily cover. A final cover of at least 60 cm (2 ft) of compacted soil is required. There are two basic methods of landfilling: area method, and trench method. Another method, "above-grade containment method," is also being used these days under special cases. These methods are discussed below.

3.6.1 AREA AND TRENCH METHODS

In *area method*, the landfilling is operated in depression, canyon, flat or rolling terrain. The cover material is obtained from the site or imported. A bulldozer spreads and compacts the waste on the natural surface of the ground, and a scaper is used to haul the cover material at the end of the day's operation. In *trench method*, a trench is excavated, and soil is stock-

TABLE 3.4. Performance Characteristics of Landfill Equipment.[a]

Equipment	Solid Waste		Cover Material			
	Spreading	Compacting	Excavating	Spreading	Compacting	Hauling
Crawler dozer	E	G	E	E	G	NA
Crawler loader	G	G	E	G	G	NA
Rubber-tired dozer	E	G	F	G	G	NA
Rubber-tired loader	G	G	F	G	G	NA
Landfill compactor	E	E	P	G	E	NA
Scraper	NA	NA	G	E	NA	E
Dragline	NA	NA	E	F	NA	NA

[a]Basis of evaluation: Easily workable soil and cover material haul distance greater than 1,000 ft.
Rating key: E = excellent; G = good; F = fair; P = poor; NA = not applicable.
Source: Brunner and Keller (1972).

TABLE 3.5. Landfill Equipment Needs.

Solid Waste Handled (tons/8 hr.)[a]	Crawler Loader		Crawler Dozer		Rubber-Tired Loader	
	Flywheel Horsepower	Weight[b] (lb)	Flywheel Horsepower	Weight[b] (lb)	Flywheel Horsepower	Weight[b] (lb)
0–20	<70	<20,000	<80	<15,000	<100	<20,000
20–50	70 to 100	20,000 to 25,000	80 to 110	15,000 to 20,000	100 to 120	20,000 to 22,500
50–130	100 to 130	25,000 to 32,500	110 to 130	20,000 to 25,000	120 to 150	22,500 to 27,000
130–250	150 to 190	32,500 to 45,000	150 to 180	30,000 to 35,000	150 to 190	27,500 to 35,000
250–500	combination of machines		250 to 280	47,500 to 52,000	combination of machines	
500–plus						

Note: Compiled from assorted promotional material from equipment manufacturers and based on ability of one machine in stated class to spread, compact, and cover within 300 ft of working face.

[a] 1 ton = 907.2 kg or 0.9 metric ton; lb × 0.4536 = kg; horsepower × 0.746 = kW.
[b] Basic weight without bucket, blade, or other accessories.
Source: Brunner and Keller (1972).

piled for use as cover material. The depth of the trench depends on the location of the groundwater and character of the soil. The collection truck deposits its load into a trench where a bulldozer spreads and compacts it. At the end of the day, the excavated soil is used as daily cover material. Figures 3.3 and 3.4 show the details of area and trench methods of sanitary landfillings.

3.6.2 ABOVE-GRADE CONTAINMENT MOUNDS

In recent years, several municipalities in flat-lying areas of the country have chosen another landfilling strategy: to pile the solid waste into enormous heaps. The municipal solid waste is placed above the ground level. The idea possibly developed from hazardous waste management practices. These isolated mountains of trash have been named *Mount Trashmores* and have been centerpieces of planned recreational areas. Houston (1973), and Beck (1973), described Virginia Beach's Mount Trashmore as typical of this new breed of man-made mountains.

In 1966, Virginia Beach faced a solid waste disposal problem. The existing 20 hectare (50 acres) sanitary landfill had only a short lifespan remaining. The high water table, 2 to 3 m (6–10 ft) made pit excavation impractical. Extending the existing sanitary landfill upward to increase its capacity and therefore its lifespan was considered. And so, in 1966, the city began construction of a 21 m (70 ft) high hill of garbage and trash. The high density achieved by daily compaction of alternating layers of 0.5 m (1.5 ft) of trash and 0.15 m (6 inches) of soil created what was proven to be a stable configuration. Through this project low value land was converted into a 34 m (110 ft) high hill with soap box derby ramps, foothills, freshwater lakes, skateboard ramps, parking areas, concession stands, playgrounds and picnic areas. There is no evidence of adverse environmental effects from Virginia Beach's Mount Trashmore. Seven seepage points for evacuation of decomposition gases have eliminated potential build-up of methane to explosive levels. Odor is not a problem because of the special care in providing daily cover. It was so well received by the public that Virginia Beach decided to build another larger Mount Trashmore and construction began in 1971. When the project is completed in 2015, a 230 hectare (570 acres) landfill will have been converted into a recreational area which will include two artificial lakes and a 43 m (140 ft) high Mount Trashmore (Beck 1973; Brown and Anderson 1988).

Another example of above-grade mound or Mount Trashmore is the Fresh Kills landfill of New York at Staten Island. Fresh Kills is reported as the largest municipal garbage dump on the planet (Figure 3.5). At its present growth rate Fresh Kills will ultimately (by 2005) evolve into one 154 m (500 ft), one 132 m (430 ft), and several 92 m (300 ft) high moun-

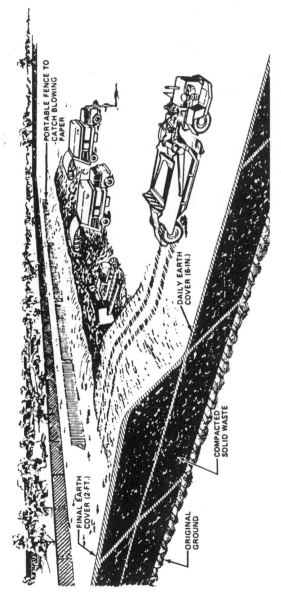

Figure 3.3 Area method of sanitary landfill operation [Source: Brunner and Keller (1972)].

PORTABLE FENCE TO CATCH BLOWING PAPER

DAILY EARTH COVER (6-IN.)

FINAL EARTH COVER (2-FT.)

COMPACTED SOLID WASTE

ORIGINAL GROUND

Figure 3.4 Sanitary landfill trench method of operation [Source: Brunner and Keller (1972)].

EARTH COVER OBTAINED BY EXCAVATION IN TRENCH

DAILY EARTH COVER (6-IN.)

COMPACTED SOLID WASTE

ORIGINAL GROUND

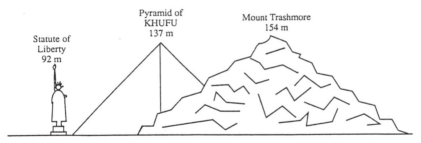

Figure 3.5 Above-grade containment mound at the Fresh Kills landfill, the world's largest landfill.

tains of trash (Brown and Anderson 1984; Swanson 1989). The landfill covers an area of almost 2,500 acres, and receives up to 14,500 tons per day of household waste. The leachate sample collection and groundwater monitoring system with over 100 wells have been reported in the literature (Anonymous 1992). Delaware and Wisconsin have also incorporated some of the above-grade mound concepts in their landfills.

Beck (1973), and Brown and Anderson (1984) gave many possible advantages of above-grade containment mounds over landfills. These are:

(1) Increased capacity per unit area of site

(2) Leachate can be removed immediately by gravitational drainage.

(3) Liners provided at the base can be sloped to reduce ponding of leachate on the liners.

(4) Repair of liners is easier. For example, if leakage becomes a problem because of the faulty cover system, it would become apparent soon after a fault develops; whereas leaking covers at below grade landfills become apparent only as a result of groundwater contamination.

(5) Solid waste is used as a construction material for building open theaters, stadiums, etc.

The disadvantages, however, include the potential for erosion and high cost because of large quantities of imported cover material. Beck (1973), and Brown and Anderson (1988) also provided the design considerations of above-grade containment mounds. These considerations are discussed below.

3.6.2.1 Site Selection

Environmentally sensitive areas as well as areas requiring complex engineering solutions (such as locations traversed by pipes) should be avoided. Also, the site selection process should include evaluation of the character and availability of on-site soil, potential socioeconomic effects of the facility, and cost estimates, taking into account future uses of the site. For environmentally sensitive areas such as wetlands, flood plains, permafrost areas, critical habitats, and recharge zones of sole source of aquifers, special care should be taken.

3.6.2.2 Geotechnical Investigations

Investigation regarding the type of soil and its bearing capacity should be carried out by digging boreholes at several locations over the site. The following parameters should be evaluated: unit weight of soil, moisture content, void ratios, angle of internal friction, and cohesion.

3.6.2.3 Geohydrological Investigations

Under this category the parameters of interest to be determined are:

- depth of the water table
- direction and rate of ground water flow
- background water quality
- current and projected ground water use
- quality, quantity, source, and seasonal variation of surface water
- potential interactions with ground and surface water
- water balance

3.6.2.4 Design Aspects

The design of a mound includes:

- an above grade containment mound, sloped to support the weight of the waste and cover
- a liner system across the base to retard entry of water and subsequent percolation of leachate
- a leachate collection and removal system that is drained freely by gravity, with drainage above ground
- a cover system consisting of a layer with gas collection equipment, a composite liner, a drainage liner, and permanent vegetative cover
- a monitoring system

3.6.2.5 Operational Aspects

The operational aspects of above-grade containment mounds include need for operating personnel and equipment costs.

3.6.2.6 Final Use

The facility designed for managing the solid waste of the city should serve after completion as a recreational area with parks, lakes, and foothills.

3.7 ENVIRONMENTAL CONCERNS

Major environmental concerns of sanitary landfill operation are paper blowing, dust, noise, traffic, odors, rodents and flies, and fires. Most of these problems can be controlled by good management, including proper compaction and covering of solid wastes. Leachate and gases are generated as the landfill ages and organic matter decomposes. The major thrust of this publication is on sanitary landfill leachate generation, control and treatment. These issues are covered in several chapters of this publication.

3.8 SANITARY LANDFILL DESIGN CONSIDERATIONS

The design goals of a sanitary landfill are to:

(1) Protect groundwater quality
(2) Protect air quality and conserve energy by installing a landfill gas recovery system.
(3) Minimize impact upon adjacent surface water and wetlands
(4) Utilize landfill space efficiently and extend site life
(5) Provide maximum use of land after completion.

Robinson (1986), Conrad et al. (1981), and Walsh and O'Leary (1991) provided a checklist of design factors for sanitary landfills. These factors are summarized in Table 3.6.

Many communities have exhausted their landfill capacities and are unable to site new facilities. New and proposed regulations are also making landfilling more difficult and expensive. Optimizing landfill design and operation will be necessary in the future to maximize the resource utilization and minimize costs. Belton (1991) provided valuable information on cost of equipment and optimization methods for landfilling operation.

Manual design of sanitary landfills is generally time consuming. In the absence of an integrated landfill design computer software, Vargas and Porter (1992) utilized computer aided drafting and design (CADD) packages at California's 360 acre *Bee Canyon Landfill* sites to make design and construction decisions and reduced design costs significantly.

3.8.1 LANDFILL LEACHATE CONTROL

Leachate control and treatment is considered a major requirement of sanitary landfill design and operation. The leachate control is achieved by various types of liners, and barrier layers applied over a specially prepared base. Five separate chapters of this book are devoted to covering the leachate issues. These chapters are Moisture Routing and Leachate Generation (Chapter 5), Leachate Characterization (Chapter 6), Leachate Attenuation (Chapter 7), Leachate Containment (Chapter 8), and Leachate Treatment (Chapter 9). Readers are referred to these chapters to obtain information on the subject of leachate control from sanitary landfills. The design features of leachate control systems are briefly presented in this section.

TABLE 3.6. **Checklist for Design of Sanitary Landfill.**

Design Factor	Description
1. Solid Waste Characterization	Determine the existing and future solid waste generation rate; and physical and chemical composition
2. Site Information	a. Perform boundary and topographic survey b. Prepare the base maps of existing conditions on and near site: property boundary, ground contours, surface waters, wetlands, utilities, roads, structures, airport, houses, other land uses c. Develop soil textural information: soil type, depth, texture, structure, bulk density, porosity, permeability, moisture content and profile, ease of excavation, stability, pH, cation exchange capacity d. Develop bedrock information: depth, type, fractures, surface outcrops e. Develop hydrological data: groundwater depth, quality, seasonal fluctuations, hydraulic gradient and direction of flow, rate of flow, current and future uses f. Identify and characterize soil cover: texture, grain size distribution, permeabilty, quantity g. Identify regulation: federal, state, local design standards, Subtitle D Regulations, local permit requirements, building codes
3. Design of Filling Area	a. Select landfilling method based on topography, site soil, bedrocks, groundwater b. Specify design dimensions: cell width, depth, length, fill depth, liner thickness, base construction and leachate collection, interim and final covers c. Specify operational features: type of soil cover, method of cover application, need for imported cover, equipment requirement, personnel requirements

TABLE 3.6. (continued).

Design Factor	Description

4. Design Features	Provide the following, if necessary:
	  a. Leachate control
	  b. Gas control
	  c. Surface water control
	  d. Access roads
	  e. Special working area
	  f. Special waste handling
	  g. Structures
	  h. Utilities
	  i. Recycling drop off
	  j. Fencing
	  k. Lighting
	  l. Washrack
	  m. Monitoring wells
	  n. Landscaping

5. Preparation of design package	a. Develop preliminary site plan of fill area b. Develop landfill contour plan: excavation, sequential fill, completed fill, fire, litter, vector, odor and noise control plans c. Compute solid waste storage volume, soil requirement volume, and site life d. Develop site plan showing: normal fill areas, special working areas, control systems for leachate, gas, surface water, access road, structures, utilities, fencing, lighting, washracks, monitoring wells, landscaping e. Prepare elevation plans with cross-section of: excavated fill, completed fill, phase development of fill at interim points f. Prepare construction details: leachate, gas, and surface water controls, access roads, structures, monitoring wells g. Prepare ultimate land use plan h. Prepare cost estimate i. Prepare design report j. Prepare environmental impact assessment k. Submit application and obtain required permit l. Prepare operator's manual m. Prepare plans for closure and post closure care

Source: Adapted in part from Conrad et al. (1981), Robinson (1986), and Walsh and O'Leary (1991).

3.8.1.1 Construction Material

Polymeric materials are used extensively to manufacture membranes and many other construction products used to control leachate from sanitary landfills and other impoundments. Among these are flexible membrane liners (FML) (also called geomembranes), geotextiles, geogrids, geonets, geocomposites, and pipes. Following is a brief discussion on the functions of each of these products (Matrecon, Inc. 1988). Koerner et al. (1989) provided a detailed discussion on these products.

3.8.1.1.1 FMLs or Geomembranes

FMLs are used to provide a barrier between mobile polluting substances released from wastes, and the groundwater. In the closing of landfills, FMLs are used to provide a low-permeability cover barrier to prevent intrusion of rain water. The main purpose of a synthetic cap is to eliminate or minimize infiltration, thus reducing leachate generation.

3.8.1.1.2 Geosynthetic Clay Liners

Geosynthetic Clay Liners (GCLs) are fabricated by distributing sodium bentonite in a uniform thickness between a woven and a non-woven geotextile. The sodium bentonite has a low permeability which makes GCLs as a suitable alternative to clay liners in a composite liner system.

3.8.1.1.3 Geotextiles

Geotextiles are used for the following purposes in landfill liners and leachate collection systems.

(1) As separation layers between two different types of soils to prevent contamination of the lower layer by the upper layer.
(2) As filters over the draining aggregates or synthetic drainage material such as geonets.
(3) As a cushion to protect synthetic liners against puncture from underlaying or overlaying rocks.
(4) As drainage layers for lateral conveyance of liquid or gases.
(5) As reinforcement for soil veneer stability, or slope stability to create steeper side slopes.

3.8.1.1.4 Geogrids

Geogrids are structural synthetic materials which are used in slope veneer stability to create stability for cover soils over synthetic liners or as soil reinforcement in steep slopes. Steeper slopers are often desired to create larger volumes for waste disposal.

3.8.1.1.5 Geonets

Geonets are synthetic drainage material which are often used in lieu of drainage sand and gravel. The advantages of geonets are that they replace

30 cm (12 inches) of drainage sand, thus increasing the landfill space for waste. They are easily installed on steep side slopes where sand and gravel would not be stable, and they provide a higher drainage uniformity than sand and gravel.

3.8.1.1.6 Geocomposites

Geocomposites are a combination of synthetic materials ordinarily used singly in geotechnical and environmental applications. The most common type of geocomposites is a composite drainage layer which consists of a geonet heat bonded or glued to two layers of geotextile, one on each side. This composite material serves as filter and drainage medium.

3.8.1.1.7 Pipes

Plastic pipes are used to provide drainage in leachate collection and removal systems and in leak-detection systems. Pipes are also used in the construction of monitoring ports, manholes, and system cleanouts.

Other materials such as sand, gravel, concrete and soils are also used to fulfill a variety of functions. Table 3.7 provides a summary of the various functions performed by various products and material.

3.8.1.2 Base Construction

The Subtitle D Regulations require installation of a composite liner system, leachate collection system, and cover system, for containing the leachate to ensure a very low groundwater carcinogenic risk level. The liner design and construction are based on specific site conditions. The earlier *natural attenuation landfills* were designed on the concept that the leachate would be attenuated (purified) by the soil beneath the landfill and by the groundwater aquifer. This concept has been discarded by most designers as serious groundwater contamination problems developed even many years after completion of landfilling. The subject of leachate attenuation is presented in Chapter 7. The new landfill design concept is based on *containment landfills*. These landfills may be single lined or double lined depending upon the site specific requirements. These requirements are applicable to new MSWLF units and later expansions. These requirements do not apply to existing units.

The final design criteria provide owners and operators with two basic design options: (1) a site-specific design that meets the performance standards and is approved by the Director of an approved State, or (2) a composite liner design. These two design options are shown in Figure 3.6.

TABLE 3.7. Materials Used in the Construction of Liner and Leachate Control Systems for Waste Storage and Disposal Facilities and Their Functions.[a]

Material	Barrier	Separation	Support	Soil Reinforcement	Filtration	Leachate Drainage and Collection[b]
FMLs	P	S	—	n/a	n/a	n/a
Geotextiles	n/a	P	—	P	P	P
Geogrids	n/a	S	—	P	n/a	n/a
Geonets	n/a	S	—	n/a	n/a	P
Composites	P or S	P or S	—	P or S	P or S	P or S
Sand/Gravel	—	S	—	—	S	P
Concrete	—	—	P	—	—	S
Pipe	—	—	—	—	—	P
Soil	P or S	—	P	—	—	—

[a]P = primary function; S = secondary function; n/a = not applicable.
[b]Also part of the leak-detection system.

Source: Matrecon, Inc. (1988)

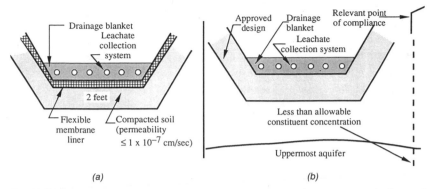

(a)   *(b)*

Figure 3.6 Design criteria for new MSWLF units and lateral expansions: (a) composite liner and leachate collection system, (b) design that meets performance standard approved by state [Source: U.S. EPA (1991)].

3.8.1.2.1 Single Lined Landfills

Single lined landfills are constructed using either clay or flexible membrane liner (FML) made of synthetic material. The selection of liner material is discussed in Chapter 8. Bagchi (1990) reported that the synthetic liners allow less leakage but are difficult to protect from damage, whereas clay liners are not easily damaged. Also, leakages through a properly constructed clay liner tend to reduce over time. Figure 3.7 shows the cross-section of a single lined landfill.

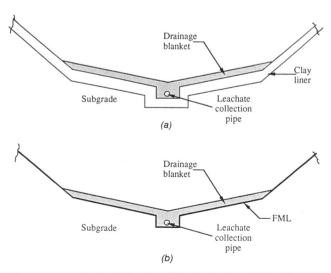

Figure 3.7 Cross-section of a single lined landfill: (a) clay liner, (b) flexible membrane liner.

3.8.1.2.2 Composite or Double Lined Landfills

More than one liner is necessary to construct a composite liner. The construction may be one FML and one clay liner or both FMLs. There may be one leachate or two leachate collection systems. The lower leachate collection (or detection) system is not expected to collect leachate unless there is a leak in the liner, or a driving head builds up over the upper liner. This may result due to blocked or insufficient flow carrying capacity of the upper leachate collection system. Several composite liners and multiple liner systems are shown in Figure 3.8.

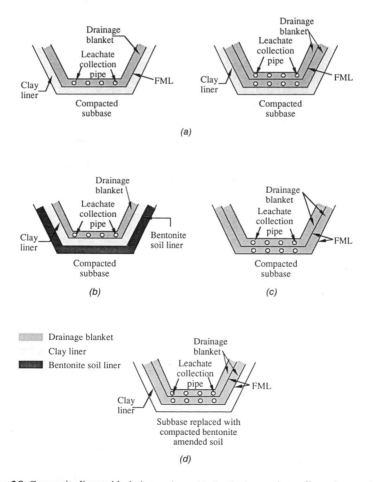

Figure 3.8 Composite liner with drainage pipes: (a) Synthetic membrane liner above and clay liner below with one and two leachate collection systems, (b) two clay liners, (c) two synthetic liners, and (d) multiple liner system.

3.8.1.2.3 Minimum Technology Guidance

EPA's minimum technological requirements for *hazardous waste landfill* design and construction required double liners and leachate collection and removal systems (U.S. EPA 1989a). Figure 3.9 is a simplified schematic diagram of a hazardous waste landfill, showing the geometry and placement of double liners and leachate collection and removal system (LCRS) in a landfill. In a double-lined landfill, there are two liners and two LCRS. The primary LCRS is located above the top liner, and the secondary LCRS is located between the two liners. In this diagram, the top liner is a flexible membrane liner (FML) and the bottom liner is a composite liner system consisting of an FML overlying compacted low permeability soil (or compacted clay). Leak detection may also be required for secondary LCRS.

3.8.1.2.4 Leachate Collection and Removal System (LCRS)

Leachate collection and removal system involves development of a grading plan, placement of drainage pipelines, and removal of accumulated leachate from the landfill. The grading plan includes sloped terraces and perforated pipes placed in each collection channel to collect and drain the leachate. Usually, the cross-slope of the terraces is 1 to 5 percent, and the slope of the drainage channel is 0.5 to 1.0 percent (Tchobanoglous et al. 1993). The leachate removal facility uses an access vault in which the leachate is collected for purifying. Design information on leachate collection, and removal systems may be found in several sources (U.S. EPA 1989a; Bagchi 1990; Tchobanoglous et al. 1993).

3.8.2 FINAL COVER

The final cover of a landfill is expected to serve many useful functions. Some of these functions have been presented in Sec. 3.2.2. It has been proposed to construct the top soil cover of several layers serving different functions. Pfeffer (1992), and Lutton et al. (1979) suggested 8-layered top cover specially designed to serve many functions. Tchobanoglous et al. (1993) proposed a modern landfill cover of five layers. U.S. EPA (1989b) in a technical guidance document on covers, provided several recommended designs. Two designs of final covers are shown in Figure 3.10. A top layer consisting of a minimum thickness of 60 cm (24 in) with 3 percent but not more than 5 percent slope is recommended [Figure 3.10(a)]. The soil drainage layer with a minimum thickness of 30 cm (12 in), and a minimum hydraulic conductivity of 1×10^{-2} cm/s will effectively minimize water infiltration into the low-permeability FML/soil layer. The minimum slope of drainage layer is 3% after settlement and subsidence. A

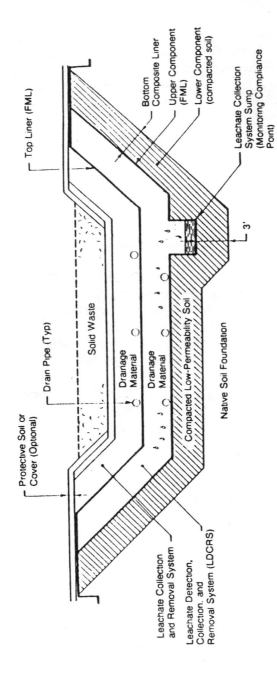

Figure 3.9 Schematic of a double liner and leachate collection system for a landfill [Source: U.S. EPA (1989a)].

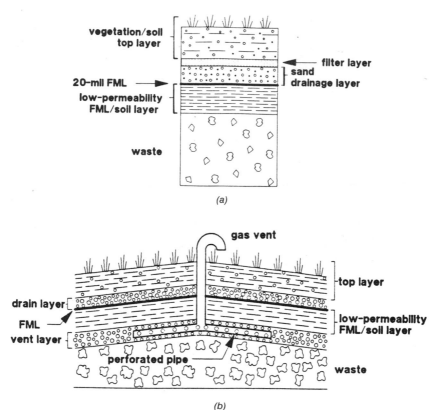

Figure 3.10 EPA recommended cover design: (a) cover with sand drainage layer, (b) cover with gas vent outlet and vent layer [Source: U.S. EPA (1989)].

two-component low-permeability layer, 20-mil FML and 60 cm (24 in) soil layer will provide hydraulic conductivity of less than 1×10^{-7} cm/s. The cover design with gas vent outlet and vent layer is shown in Figure 3.10(b). The performance of the cover under field conditions may be questionable due to differential settlement, surface subsidence, and other adverse conditions under long-term post closure care.

3.8.3 SURFACE WATER CONTROL

Surface water control is necessary to divert the rain water away from the landfill. Grading and compacting the landfill to a profile of 6 to 12 percent will allow surface water to drain from the surface. Diversion ditches are constructed around the perimeter of the filled area for diversion of upland rainfall, as well as collection of the surface runoff from the landfill site.

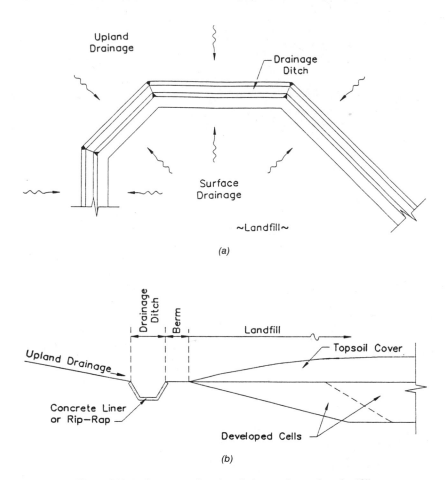

Figure 3.11 Surface water diversion ditches used at sanitary landfills.

The diversion ditches are generally designed for a 10-year storm intensity. Diversion ditches are usually trapezoidal, semicircular, or vee shaped. These ditches are also generally concrete lined. Figure 3.11 shows details of a type of surface water diversion ditch used in sanitary landfills.

3.8.4 SANITARY LANDFILL GAS CONTROL AND COLLECTION

3.8.4.1 Gas Generation

Landfill gas is composed primarily of equal parts methane and carbon dioxide, with trace organic chemicals (e.g., benzene, trichloroethylene, vinyl chloride, methylene chloride) also present (U.S. Congress 1989).

Methane production begins once conditions in a landfill become anaerobic. Rates of methane production depend on the moisture content of the landfill; concentrations of nutrients and bacteria; pH, age, volume and density of solid waste, and the presence or absence of sewage sludge. Knox (1992) provided information on gas production rate from a sanitary landfill. Typical gas generation rates range from 3 to 13 L/kg per year (0.05 to 0.20 ft³/lb year). Another method of estimating is 400 to 600 L of gas per kg of solid waste over the lifetime of the landfill (6.5 to 9.6 ft³/lb). The duration of gas generation, however, is difficult to calculate. Site histories indicate that gas generation continues for at least 10 or even 15 years (Knox 1992). The heat value of typical landfill gas is about 18,500 kJ per sm³ (500 BTU/sft³). The energy content of the gas varies depending upon the performance of the gas collection system and the phase of decomposition. To raise the heat content of the gas, removal of moisture, carbon dioxide, particulate matter, and trace contaminants will be required. For comparison, the natural gas has heat content of about 37,000 kJ/sm³ (1,000 BTU/scf). Purification involves using chemical and physical processes such as dehydration by trimethylene glycol process, molecular sieves and refrigeration, etc. (Robinson 1986; U.S. EPA 1988a, 1988b).

3.8.4.2 Gas Collection Systems

Gas collection systems are constructed to manage and control the movements of landfill gas. These may be passive or active systems. A passive system functions on the principle of natural pressures and convection mechanisms to move the landfill gas. Active systems move the landfill gas under an induced negative pressure or vacuum (Robinson 1986; Knox 1992).

Passive systems utilize barriers, gravel-filled trenches, pipes and vents. These systems have the advantage of lower operating and maintenance costs. There must, however, be a natural pressure gradient within the landfill for passive systems to function. Knox (1992), and Tchobanoglous et al. (1993) reported that such systems are not suitable where methane accumulation poses a significant hazard. The passive gas venting and recovery systems are shown in Figure 3.12. The gases are vented, flared, or collected.

The active landfill gas collection system is the most effective means for collecting landfill gas, while permeable trenches are the least effective (U.S. EPA 1988). In active landfill gas control systems, the gas is collected through vertical or horizontal extraction wells to collection headers; from there the gas is driven by a vacuum compressor or blower to a collection area to be flared or recovered (Flanagan 1988; U.S. Congress 1989; Knox 1992; Pfeffer 1992; Tchobanoglous et al. 1993). The extraction wells typi-

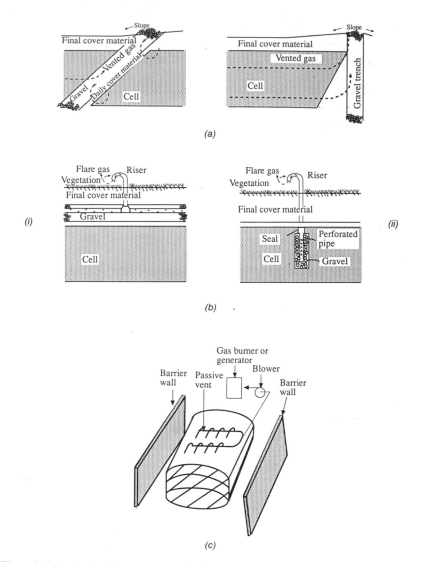

Figure 3.12 Passive landfill gas venting and collection systems: (a) gravel vents and gravel filled trench, (b) vent pipe, (i) above refuse cell, (ii) inside refuse cell, (c) gas recovery.

cally consist of a perforated pipe casing surrounded by permeable material such as gravel, and an impermeable seal near the top, placed in a borehole made through the refuse. Knox (1992) reported that extraction boreholes are typically 30 to 91 cm in diameter (12 to 36 in) and the extraction well is typically 15 to 30 cm in diameter (6 to 12 in). It is important to provide an airtight seal at the top to prevent air from draining into the system. Excessive vacuum may also draw air. The collection header pipes connect the vertical gas extraction well with the blower. The gas is routed through these pipes to a central location where it is flared or used for power generation. Pfeffer (1992) described the active landfill gas recovery system. This system is shown in Figure 3.13. Flanagan (1988), reported that these systems are operated only in portions of landfills that have been closed temporarily or permanently with a cap, although they can be installed as the fill is built and then later connected.

Because of the moisture content and temperature of the environment, the produced gas is nearly saturated. Therefore, condensate is probably the greatest single cause of problems in the collection system. Condensate traps are used to remove the condensate from the collection header pipes. Condensate quantity is generally 20 to 80 L per million L of gas. The typical spacing of condensate trap in the header pipe is 60 to 200 m (Knox 1992).

Subtitle D regulations require that concentration of methane generated by municipal solid waste landfills not exceed 25 percent of the lower explosive limit (LEL) in on-site structures, or the LEL at the facility property boundary (U.S. EPA 1991b). The U.S. EPA has proposed controls for air emissions from municipal solid waste landfills based on Section III of the Clean Air Act. This proposal was published in the Federal Register on May 30, 1991.

If methane emissions were collected completely and processed for energy recovery, they could significantly reduce natural gas consumption. In the United States, only about 123 landfills actually collect methane to recover energy (U.S. EPA, 1988a and 1988b). The collected gas can be purified to increase its energy, and then be used in boilers, space heaters, and turbines. Feit (1991) discussed landfill gas utilization technologies for power generation.

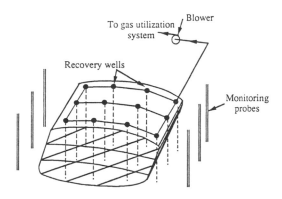

Figure 3.13 Active landfill gas recovery system.

3.9 LANDFILL STABILIZATION TECHNIQUES

The principal difficulties in adapting the method of sanitary landfilling to solid waste disposal is the fact that the landfill site may not easily be used until the solid wastes have decomposed within the fills, the production of gases has ceased, and the settlement of the fill surface has stabilized. In order to reduce the difficulties associated with the long decomposition times in sanitary landfills, several techniques have been proposed for acceleration of the biodegradation of solid wastes. These techniques include incineration of the wastes within fill cells, forced aeration of the wastes within the cells to promote rapid aerobic degradation, and recycling of leachate produced by abundant watering of the landfill surface to create, in effect, a solid waste treatment plant. These techniques are presented below.

3.9.1 TESTS OF IN-SITU LANDFILL INCINERATION

In a series of controlled landfilling operations in the state of California, the Ralph Stone Company (1969) under contract with the Office of Solid Waste Management Programs, has tested the technical feasibility of placing solid wastes in conventional sanitary landfill cells, and then initiating drying of the wastes by means of heating elements placed within them, further heating the refuse to the ignition point, and allowing a contained combustion to take place. In a series of preliminary tests, the operation required a period of more than one week for the drying and combustion of the wastes in a typical cell. Drying of the wastes to the ignition point required several days, and after ignition the primary combustion of most of the wastes required an additional period of several days. Finally, large, bulky, slow-burning items, such as sizable pieces of wood or logs burned for one or two days after the major portion of the waste had been consumed. On the basis of preliminary tests, the investigating personnel concluded that such means of stabilization of a landfill is technically feasible. The economics of such stabilization have not been proven to date. Ralph Stone and Company (1969) provided the test results, and recommendations.

3.9.2 FORCED AERATION

Ralph Stone and Company (1969), also investigated the concept of forced aeration of landfill cells by means of a buried piping system in the field. In this method, excess air was supplied to the solid wastes within a landfill cell and the waste gases produced during decomposition were removed so that essentially aerobic decomposition took place within the fill. The maintenance of aerobic degradation within the fill cells accelerated the decomposition of the wastes and did not produce the explosive or toxic gases associated with anaerobic decay. As in the case of subsurface incineration, the investigating personnel concluded that aerobic degradation of wastes within a landfill by means of forced aeration is technically feasible.

3.9.3 LEACHATE RECYCLE

Some researchers and reports suggest that the decomposition process could be enhanced by collecting leachate and recycling it back into the organic material, an idea that has been examined in laboratory situations. (Tittlebaum 1982; Harper and Pohland 1988a, 1988b; U.S. EPA 1988; U.S. Congress 1989). Recycling leachate in some manner is used in many landfills today.

One study at a Pennsylvania landfill concluded that recycling leachate resulted in more rapid decomposition, enhanced methane production, and increased stabilization (U.S. EPA 1988a, 1988b). The Delaware Solid Waste Authority recently initiated a study to examine this idea on a larger scale under field conditions. Two 0.5 hectare (1-acre) landfill cells for household solid wastes only were set up; leachate was collected and removed for external disposal from one cell, and recycled in the other cell. Decomposition rates were measured after 5 years (U.S. Congress, 1989). Several other studies were conducted to determine the feasibility of landfill stabilization. Additional detail may be found in Chapters 6 and 9.

Experimental data on leachate recycle indicate several potential benefits:

- The time needed to decompose organic materials might be reduced from around 15 years to only a few years.
- Methane production could be maximized, making recovery more viable.
- Reusable space would become available more rapidly.
- Landfill could be used as an equilization basin.
- Collected leachate will have lower biodegradable organics to be treated at wastewater treatment plants.

However, the researchers have also noted several problems:

- Uncertainties exist about the ultimate reactivity or fate of the chemical compounds created during the process.
- Regulatory proposals by EPA may ban the addition of any liquids to landfills.
- These designs would require careful design and location of the leachate collection system in the landfill cell, and controlled rates

of leachate recycling to minimize potential for clogging and off-site migration.

Harper and Pohland (1988) have noted that the increased volume of leachate may clog the leachate collection system. Also, small tears in the liner during construction or daily operation may cause leachate migration due to recycling. These potential problems suggest that enhanced decomposition be used only at sites that are not located near groundwater.

3.10 RECYCLING LANDFILL SPACE

"Recycling" landfills is reusing the same landfill space after a period of decomposition. This has been suggested as a means of reusing landfill space, allowing the repair of liners and leachate collection systems, and recovering materials of value (Knapp and Sulton 1981; Vasuki 1988). One recycling operation in Collier County, Florida, mined a municipal landfill and processed materials at a centralized facility. Screening was used to remove fine soil and humus and to recover ferrous metals (Gershman, Brickner, and Bratton, Inc., 1988).

The Delaware Solid Waste Authority plans to excavate a 3 hectare cell containing municipal solid wastes deposited between 1980 and 1982. The excavation will start when degradation is essentially over, as measured by decreased methane production. The excavated material will be screened for ferrous scrap, plastics, wood, textiles, aluminum, glass, and other materials; some of these could be burned (to recover energy) while others could possibly be recycled. The Authority intends to rebuild the liners and leachate collection systems so that the area can be reused, and to use the screened dirt as daily cover material (U.S. Congress 1989).

3.11 CLOSURE AND POST-CLOSURE CARE

Subtitle D regulations clearly spell out the closure and post-closure care of MSWLFs (Chapter 4). After completion of landfilling activity, the site must be closed properly and post-closure care is continued in accordance with federal or state regulations. In this section the major elements of closure, post-closure care and liability and financial assurance are presented. Figure 3.14 shows the completion plan of a sanitary landfill.

3.11.1 DEVELOPMENT OF A CLOSURE PLAN

Most of the states require development of a closure plan before the construction and landfilling operation is started. The closure plan must pro-

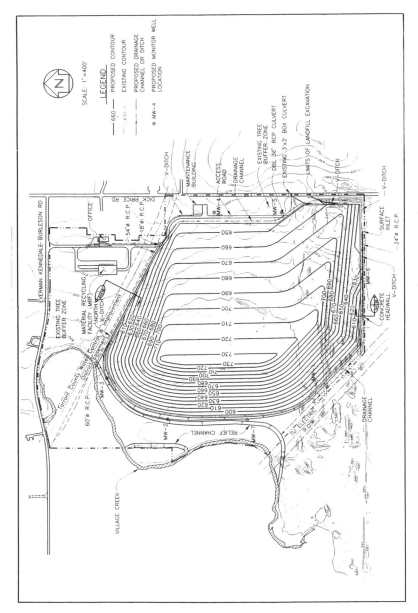

Figure 3.14 Completion plan of a sanitary landfill (courtesy Waste Management Inc., and Freese and Nichols, Consulting Engineers).

vide information on cover design, ultimate use, monitoring needs of environmental factors (air, water, land), cost estimates, financial assurance for post-closure care over the specified period, corrective actions, and the like. During the active life the information may be revised and updated. A final closure plan is prepared and adapted just before the landfill stops receiving the solid wastes.

3.11.2 ELEMENTS OF A CLOSURE PLAN

There are five major elements of a closure plan that must be addressed. These are:

(1) Final cover design
(2) Surface water control
(3) Control of gases
(4) Leachate control
(5) Environmental control

Each of these elements are briefly discussed below.

3.11.2.1 Final Cover Design

The final cover should be designed to serve many basic functions discussed in Secs. 3.3.2 and 3.8.2. Tchobanoglous et al. (1993) provided six design parameters of a typical final cover. These are: (1) design configuration, (2) final permeability, (3) surface slope, (4) landscape design, (5) method of repair as the landfill settles, and (6) slope stability under static and dynamic loadings. If the final use requires a green area, a revegetation plan utilizing native plants should be included. The final cover design and construction should also include construction quality assurance program (CQA), and consideration of engineering properties of soils, drainage grid, and synthetic membranes if required. Tchobanoglous et al. (1993) also provided an excellent discussion on revegetation of closed landfill sites.

3.11.2.2 Surface Water Control Systems

Both run-on and runoff of surface water should be controlled by providing diversion channels around the perimeter of the landfill. Proper grading and compaction of the final cover is also necessary. These topics have been fully discussed in Sec. 3.8.3. A plan for surface grading is necessary to repair the ponding that may develop due to uneven settlement.

3.11.2.3 Control of Gases

Landfill gases will start generating soon after the landfill is completed, and will continue to be generated for a long time. If not properly controlled, the gases will migrate in the soil and be a source of hazard for buildings nearby. Gas collection systems are covered in Sec. 3.8.4. A plan to check, and properly dispose of or to utilize the gases should be covered in the closure plan.

3.11.2.4 Leachate Control

Leachate migration from a sanitary landfill constitutes the biggest threat to groundwater pollution. The leachate must be contained, collected and treated. Impervious liners are required to contain and remove the leachate from the landfill as it is generated. Types and design of liners are presented in Sec. 3.8.1. Chapter 8 is devoted to leachate containment and liner selection. Once the leachate is contained and collected it must be treated prior to disposal. Various options for disposal are: pretreatment prior to discharge into POTW, solar evaporation, land application, and complete treatment prior to discharge into receiving waters. Chapter 9 is devoted entirely on leachate treatment and disposal. The closure plan should provide a discussion on the leachate treatment facility, an upgrading plan as the leachate quality changes with time, and operation and maintenance strategy.

3.11.2.5 Environmental Control

Environmental control requires close monitoring of parameters that are indicators of the quality of water, air, and land.

3.11.2.5.1 Water

Groundwater monitoring in the uppermost aquifer, and soil under the landfill must be done to detect any leakage that may develop in the impervious liner system. The soil monitoring facility must be capable of functioning in the *vadose zone* because in dry climates, moisture may not penetrate to soils beneath the landfill. A sufficient number of groundwater monitoring wells must be provided to obtain samples from the uppermost aquifer upstream, downstream, around the boundary, and through the landfills. All wells must be capped to insure the integrity of the borehole. If groundwater contamination is detected, corrective actions must be taken in accordance with the Subtitle D requirements (Chapter 4).

As an example of groundwater monitoring, in the world's largest landfill site at Fresh Kills, over 100 wells were originally constructed in the first phase. An additional 120 wells have been added in the second phase. Many innovative ideas have been used in the pump selection that reduce the volume of water for priming and purging the lines before sampling. Other ideas involve well caps, and inflatable packer devices to fill and seal the inside of the well casing (Anonymous, 1992).

3.11.2.5.2 Air

The landfill closure plan must show the manner in which methane and other gases are controlled. The gas samples collected must be analyzed periodically to assess the degree of biological activity in the landfill.

3.11.2.5.3 Soil

Environmental monitoring of soil include measuring surface settlement, soil slippage and land surface erosion. The final cover placement must be under strict construction supervision. The cover must be inspected frequently and maintained. Proper surface slope and filling of potholes is necessary to minimize percolation. Vegetation as originally planned must be maintained.

3.11.3 POST-CLOSURE CARE

The landfill, after closure, must be maintained over a period of 30 years. Post-closure care involves maintaining the integrity and effectiveness of the final cover, operating the leachate-collection system, monitoring the groundwater, and maintaining and operating the gas collection system. Post-closure care must allow for the planned uses of the site. The name, address and telephone number of the person in office to be contacted about the facility during the post-closure period must also be posted.

The post-closure care requirements are met by routine inspection. A well defined maintenance plan must be set in place for grading and landscaping, surface drainage control, gas monitoring and control, and leachate collection and treatment.

3.11.4 LIABILITY AND FINANCIAL ASSURANCE

Post-closure care and liability become less clear as the closed landfill site is sold or resold. The responsibilities are: (1) as the owner of the site, and (2) liabilities for the waste in it. If the site is owned by a municipality the responsibility is in both areas. If however, the owner is independent of the waste generator, the responsibility is split between the owner and generator. The laws require the site owner to operate, close and provide post-closure care of the landfill in accordance with the approved guidelines. Financial assurance is required by the owner or operator for post-closure care and corrective actions. The estimates for post-closure care are developed by hiring a third party to perform closure and post-closure care over the specified period. The allowable financial assurance mechanisms are stated in Subtitle D. Readers may refer to Chapter 4 for additional information on Subtitle D requirements.

3.12 EXAMPLE OF A MODERN LANDFILL

Santama Landfill, located in Tokyo, is an example of a modern landfill. Although it is not representative of most Japanese landfills, it is likely to be representative of many future designs. The following description has been adopted from a report (U.S. Congress 1989).

The landfill is a joint venture among 25 municipalities and 2 towns. Funds for operating the facility come from taxing the contributing municipalities and charging a tipping fee. The site was identified in 1981, and although some public opposition was encountered, negotiations lead to an agreement that the facility would utilize advanced pollution control techniques. About 22 hectares will be used for landfilling, with a surrounding undeveloped green zone of 14 hectares. It came on line in 1984, and at the end of its useful life (expected to be about 13 years) the site will be capped and transformed into a sports facility. Details of post-closure monitoring and leachate collection and treatment plans are not known.

One unique aspect of the Santama facility is that, like many Japanese landfills, it does not accept organic wastes (paper, food, yard wastes). Instead, these wastes are collected by the municipalities and sent to incinerators. It also does not accept industrial wastes. It does accept fly and bottom ash from the municipal wastes incinerators (mostly untreated but about 10 percent processed by cementation).

Another unique aspect is its inclusion of many different engineering features: computerized weigh-in, record-keeping system for each truck, truck washing system, intricate liner system and "sandwich" process, leachate collection and drainage pipes, groundwater flow channels, secondary wastewater treatment plant for leachate, groundwater monitoring wells, and gas venting. The bottom liner system consists of 1 meter of thick clay covered by a synthetic rubber liner and then another 1 meter of clay. The filling of the landfill is based on a "sandwich process" — each 2-meter layer of solid waste is covered with a 1 meter thick clay lining. The wastewater treatment facility provides activated sludge secondary treatment for the leachate. It removes an average of 92 percent of the bio-chemical oxygen demand (BOD), bringing post-treatment levels down to 8 to 10 mg/L. The leachate is then disinfected with chlorine and sent to a sewage treatment plant in Tokyo. BOD is tested weekly at an on-site laboratory; cyanide, PCBs, nitrates, phosphorus, and seven metals are tested monthly.

The design is not without some problems and controversies. The landfill is located along a small stream, which dictates the need for a disaster prevention flood control system. One reviewer also suggests that the sandwich process may waste space and inhibit internal drainage. Data are not available on leachate volume and characteristics or on hydrological balance, so it is difficult to evaluate the effect of this feature.

3.13 REFERENCES

Anonymous. 1992."How Largest Landfill Samples Groundwater," *Water Engineering and Management,* 139(4):29.

Bagchi, A. 1990. *Design Construction, and Monitoring of Sanitary Landfills,* New York: John Wiley and Sons.

Beck, W. M., Jr. 1973. "Building an Amphitheater and Coasting Ramp of Municipal Solid Wastes," Vols. I and II, PB 225 346, *NTIS,* U.S. Environmental Protection Agency.

Belton, N. 1991. "Optimizing Your Landfill Operation," *Waste Age,* 22(9):93–96.

Brown, K. W. and D. C. Anderson. 1988. "Above-Ground Disposal," in *Standard Handbook of Hazardous Wastes Treatment and Disposal,* H. M. Freeman, ed., New York: McGraw-Hill Book Co., pp. 10.85–10.91.

Brown, K. W. and D. C. Anderson. 1984. "Above-Ground Storage of Waste," Conference and Exhibition on Hazardous Waste and Environmental Emergencies, Houston, Texas, March 12–14.

Brunner, D. R. and D. J. Keller. 1972. "Sanitary Landfill Design and Operation," U.S. Environmental Protection Agency, Solid Waste Management Series SW-65ts.

Conrad, E. T., J. J. Walsh, J. Atcheson and R. B. Garner. 1981. "Solid Waste Landfill Design and Operation Practices," EPA Draft Report Contract No. 68-01-3915.

Davis, M. L. and D. A. Cornwell. 1991. "Introduction to Environmental Engineering," Second Edition, New York: McGraw-Hill, Inc.

Feit, E. 1991. "Turning Landfill Gas into Kilowatts," *Waste Age,* 22(3):130–134.

Flanagan, K. 1988. "Methane Recovery Does More Than Provide Energy," *Solid Waste and Power,* 2(4):30–33.

Gersham, Brickner and Bratton, Inc. 1988. "Performance, Constraints, and Costs of Municipal Solid Wastes Management Technologies," Contract Report Prepared for U.S. Congress, Office of Technology Assessment, Falls Church, Va.

Harper, S. R. and F. G. Pohland. 1988. "Landfill Lessening Environmental Impacts," *Civil Engineering,* 58(11):66–69.

Houston, J. 1973. "The Rise of Mount Trashmore," *Park and Recreation,* 8(1):28–30.

Knapp, D. and C. Sulton. 1981. "On the Recycling of Landfills," *Materials World Publishing,* Berkeley, California.

Knox, T. D. 1992. "Landfill Gas Management Technologies," *Waste Age,* 23(4):233–244.

Koerner, R. E., Y. Halse-Hsuan, and A. R. Lord. 1991. "Long-Term Durability of Geomembranes," *Civil Engineering,* ASCE, 61(4):56–58.

Lutton, R. J., G. L. Regan, and L. W. Jones. 1979. "Design and Construction of Covers for Solid Waste Landfills," EPA-600/2-79-165, *U.S. Environmental Protection Agency,* Cincinnati, Ohio: Municipal Environmental Research Laboratory.

Matrecon, Inc. 1988. "Lining of Waste Impoundment and Other Impoundment Facilities," EPA/600/2-88/052, *U.S. Environmental Protection Agency,* Cincinnati, Ohio: Risk Reduction Engineering Laboratory, Office of Research and Development.

Meyer, M. P. and S. J. Knight. 1961. "Trafficability of Soils, Soil Classification," U.S. Army Engineer Waterways Experiment Station, Vicksburg, Mississippi: Technical Memorandum 3-240, Supplement 16.

Noble, A. 1976. "Sanitary Landfill Design Handbook," *Technomics,* Westport, Connecticut, p. 285.

O'Leary, P., B. Tansel, and R. Fero. 1986. "How to Evaluate a Potential Sanitary Landfill Site," *Waste Age,* 17:78–99.

Pfeffer, J. T. 1992. *Solid Waste Management Engineering,* Englewood Cliffs, N.J.: Prentice Hall.

Ralph Stone and Co. 1969. "Solid Waste Landfill Stabilization," An Interim Report to the Office of Solid Waste Management Program, USPHS.

Robinson, W. D. 1986. *The Solid Waste Handbook. A Practical Guide,* New York: John Wiley and Sons.

Sorg, T. J. and H. L. Hickman. 1970. "Sanitary Landfill Facts," SW-4ts, U.S. Department of Health, Education and Welfare, Public Health Service, Washington, D.C.

Swanson, D. J. 1989. "New York Mounts Recycling Campaign as Colossal Garbage Heap Grows Sky-High," *Dallas Morning News Sunday,* July 30, pp. 26A.

Tchobanoglous, G., H. Theisen, and S. Vigil. 1993. "Integrated Solid Waste Management, Engineering Principles and Management Issues," Second Edition, New York: McGraw-Hill.

Tittlebaum, M. D. 1982. "Organic Carbon Content Stabilization Through Landfill Leachate Recirculation," *Journal Water Pollution Control Federation,* 54(5):428–433.

U.S. Congress. 1989. "Facing America's Trash: What Next for Municipal Solid Wastes," OTA-0-424, *U.S. Government Printing Office,* Washington, D.C.

U.S. Environmental Protection Agency. 1988a. "Criteria for Municipal Solid Waste Landfills," EPA/530-SW-88-037, Office of Solid Waste, Operating Criteria (Subpart C), Washington, D.C.

U.S. Environmental Protection Agency. 1988b. "Solid Waste Disposal in the United States," Vol. II, EPA/530-SW-88-011B, Report to Congress, Washington, D.C.

U.S. Environmental Protection Agency. 1989a. "Requirements for Hazardous Waste Landfill Design, Construction and Closure," Seminar Publication, EPA/625/4-89/522, *Technology Transfer,* Cincinnati, Ohio.

U.S. Environmental Protection Agency. 1989b. "Technical Guidance Document: Final

Covers on Hazardous Waste Landfills and Surface Impoundments," EPA/530-SW-89-047, *Office of Solid Waste and Emergency Response,* Washington, D.C.

U.S. Environmental Protection Agency. 1991a. "Solid Waste Disposal Facility Criteria; Final Rule," *40 CFR Part II,* 257 and 258, Subtitle D of the Resource Conservation and Recovery Act (RCRA).

U.S. Environmental Protection Agency. 1991b. "Explosive Gases Control, Operating Criteria," *40 CFR Part 258.23, Subpart C,* Operating Criteria, Explosive Gases Control.

Vasuki, N. C. 1988. "Why Not Recycle the Landfill," *Waste Age,* 19(11):165–170.

Vargas, J. C. and D. B. Porter. 1992. "Landfill's: Anatomy of Automated Design," *Civil Engineering, ASCE,* 62(3):52–55.

Walsh, P. and P. O'Leary. 1991. "Sanitary Landfill Design Procedures," *Waste Age,* 22(9):97–105.

CHAPTER 4

Subtitle D Regulations and Their Impacts

4.1 INTRODUCTION

SOLID waste related legislation in the United States dates from 1965, when the Solid Waste Disposal Act, Title II of Public Law 89-272 was enacted by Congress. Since then, major legislative actions have been taken in the solid waste area, to encourage resource recovery and protect the groundwater quality. On September 11, 1991, the U.S. Environmental Protection Agency promulgated its long-awaited Subtitle D regulations for municipal solid waste landfills. Both existing and future landfills will be significantly affected by these regulations.

The purpose of this chapter is to briefly review the major legislations that relate to solid waste disposal activities. Since Subtitle D regulations will dictate the future design and operation of municipal landfills, these regulations, and the impact of these regulations upon current and future landfilling operations, are presented in greater detail.

4.2 SOLID AND HAZARDOUS WASTE LEGISLATION

In 1965, the U.S. Congress passed the first national solid waste legislation, and the Solid Waste Disposal Act of 1965 (P.L. 89-272) was enacted (U.S. Congress 1965). This Act provided money to the states to develop solid waste management plans and to survey current disposal practices. In 1970, the Resource Recovery Act (P.L. 91-512) was passed (U.S. Congress 1970). This Act marked a significant policy change from focusing on disposal problems to examining recovery processes for materials and energy. In October 1976, the Resource Conservation and Recovery Act (RCRA) was enacted (U.S. Congress 1976a). This Act restricts the disposal of hazardous waste into land, bodies of water, or the atmosphere. Thus, RCRA was enacted to protect the quality of groundwater, surface water, the land and the air from contamination by solid wastes.

The Toxic Substance Control Act of 1976 (TSCA) was enacted to regulate the introduction and use of new hazardous chemicals (U.S. Congress 1976b). Under TSCA regulations, industry must furnish data on the anticipated production, usage, and health effects of all new chemical substances and mixtures, before they are manufactured for commercial distribution. TSCA also regulates the manufacture, processing, use, and final disposal of all chemical substances.

The Hazardous and Solid Waste Amendments of 1984 (HSWA) emphasized the protection of groundwater through the use of leachate collection (double liners) and monitoring of underground tanks, upgraded criteria for disposing of municipal solid wastes in landfills, and established new requirements for the management and treatment of small quantities of hazardous wastes (U.S. Congress 1984).

The Comprehensive Environmental Response Compensation and Liabilities Act of 1980 (CERCLA), the so-called *Superfund Legislation,* was designed primarily to address the problem of financing the cleanup of abandoned or illegal hazardous waste sites (U.S. Congress 1980). The Superfund Amendments and Reauthorization Act of 1986 (SARA) provided funds, a timetable and guidance for cleanup standards (U.S. Congress 1986).

4.3 RCRA SUBTITLE D

In the Federal Register of October 9, 1991, the U.S. Environmental Protection Agency published *40 CFR Parts 257 and 258, Solid Waste Disposal Facility Criteria; Final Rule* (U.S. EPA 1991). These parts included a revision to 257, and Regulations in Criteria for Classification of Solid Waste Disposal Facilities and Practices. The final version of Part 258 – Criteria for Municipal Solid Waste Landfills (MSWLFs) was signed on September 11, 1991. The subparts of Part 258 deal with the following issues:

- Subpart A – General
- Subpart B – Location Restrictions
- Subpart C – Operating Criteria
- Subpart D – Design Criteria
- Subpart E – Groundwater Monitoring and Corrective Actions
- Subpart F – Closure and Post Closure
- Subpart G – Financial Assurance Criteria

Known as the "Subtitle D Rule," Part 258 establishes minimum federal criteria for MSWLFs. These regulations address location restrictions, facility design criteria, operation criteria, groundwater monitoring requirements, corrective action requirements, financial assurance requirements, and closure and post closure care requirements. Repa (1991), and Pfeffer (1992) provided a summary of Subtitle D regulations. In accordance with these criteria, landfill owners and operators are required to take steps to comply with the new requirements by October 9, 1993. Both existing and new landfills are regulated under these rules.

4.3.1 SITE REQUIREMENTS

A sanitary landfill that is currently open and accepting solid waste as of October 9, 1991 can continue to operate unless it is located on one of the three types of sites EPA considers high-risk areas for landfilling. These high risk areas are (a) airports, (b) 100-year floodplains, and (c) unstable areas. Landfills falling within these areas may be subject to closing. New landfills and expansions to existing landfills must also comply with additional site restrictions for wetlands, fault areas, and seismic impact zones. Some of the requirements of Subtitle D are summarized below:

(1) *Airports*—A municipal solid waste landfill located near an airport will have to demonstrate that it does not pose a bird hazard to aircraft. Landfills in the following locations will be affected: (a) within 3048 m (10,000 ft) of the end of any airport runway used by turbojet aircraft, and (b) within 1574 m (5,000 ft) of the end of any airport runway used by piston-type aircraft.

(2) *Floodplains*—A landfill located within a 100-year floodplain will have to demonstrate that it does not (a) restrict the flow of the 100-year flood, (b) reduce the temporary water storage capacity of the floodplain, or (c) contribute to the washout of solid wastes, which would pose a hazard to human health or the environment.

(3) *Wetlands*—A landfill can only operate in wetlands if the owner or operator can meet the following requirements: (a) no significant environmental consequences, (b) no violations of state water quality standards, (c) no violation of the *Clean Water Act*, (d) no violation of the *Marine Protection Research and Sanctuary Act* of 1972 (U.S. Congress 1972), (e) no degradation of wetlands, (f) landfill design must minimize potential adverse impacts, and (g) sufficient information must be available to make soil determinations.

(4) *Fault Areas*—New units of landfill cannot be located within 66 m (200 ft) of a fault that has had any displacement.

(5) *Seismic Impact Zones*—All structures, liners, leachate collection systems, and subsurface water control systems of new landfills located in a "Seismic Impact Zone," must be designed to resist the horizontal acceleration resulting from seismic activity in the area.

(6) *Unstable Areas*—Landfills located in an unstable area must demonstrate structural stability. Unstable areas are locations with a high potential for groundwater or soil movement at or surrounding the landfill site including (a) uneven support caused by poor foundation conditions, (b) landslides, avalanches, and (c) sinkholes, sinking streams, caves, large springs, and blind valleys.

4.3.2 OPERATING CRITERIA

The Subtitle D regulations define the following operating criteria:

(1) *Detection/Exclusion of Hazardous Waste*—The owner or operator must implement a program to detect and exclude the disposal of regulated solid waste and waste containing polychlorinated biphenyls (PCBs). The program must include provisions to inspect incoming loads, to record such inspections, to train personnel to recognize regulated hazardous waste, and to notify the proper authorities of detected hazardous wastes. Hazardous wastes are those that exhibit *ignitability, corrosivity, reactivity or toxicity*.

(2) *Cover Material*—The cover material must be applied at least daily to minimize vectors, fires, odors, blowing litter, and scavenging.

(3) *Disease Vector Control*—The owner or operator must prevent or control on-site population of disease vectors such as flies, mosquitoes, or other animals and insects.

(4) *Explosive Gases Control*—The concentration of methane must not exceed more than 25% of the lower explosive limit (LEL) in any facility structure, or 100% LEL at the facility property boundary. The facility must conduct routine methane monitoring to comply with the requirements governing methane buildup. This monitoring frequency must be at least quarterly. There are provisions in Subtitle D which state exactly what must be accomplished if the levels stated above are exceeded.

(5) *Air Criteria*—A landfill cannot violate any requirements developed under a *State Implementation Plan* (SIP) to comply with the *Clean Air Act*. Open burning of solid waste is prohibited except for infrequent burning of agricultural and silvicultural wastes.

(6) *Access*—A landfill must control public access to prevent unauthorized traffic and illegal dumping.

(7) *Run-on/Run-off*—The design of the landfill must prevent run-on and run-off during the peak discharge from the 25 year flood. Also, the active portions of the landfill must be able to collect and control at least the water volume from a 24-hour, 25-year storm.

(8) *Surface Water*—The landfill must not violate provisions of the *Clean Water Act* including discharging pollutants from point source and non-point discharges.

(9) *Liquids Restrictions*—The acceptance of liquids in a landfill is limited, but household liquid waste can be accepted. If the facility is equipped with a composite liner and a leachate collection system, then the facility may recirculate leachate.

(10) *Codisposal of sludge in MSWLFs*—In many states codisposal of municipal wastewater and water treatment plant sludges in MSWLFs is practiced. The requirements are: (a) a maximum moisture content in the sludge, and (b) a maximum ratio of sludge to municipal solid waste.

The Final Rule of the EPA amends regulations under 40 CFR 257 and 403, and adopts a new Part 40 CFR 503 to establish requirements for the final use and disposal of sewage sludge generated by the treatment of domestic sewage and municipal wastewater in publicly or privately owned treatment works (POTW). Sludge that is co-disposed with municipal solid waste in a sanitary landfill will not have to meet numerical pollutant limits as with other disposal/beneficial use practices regulated under 40 CFR 503 (U.S. EPA 1993). However, sludges must not be hazardous and must not contain free liquids. EPA has not established numerical pollutant limits for wastewater sludge co-disposed with municipal solid waste, and it is difficult to predict how pollutants in the co-disposed sludge will move from a landfill into the environment. However, EPA believes that other environmental protection requirements for landfills such as liner protection and groundwater monitoring under Subtitle D are sufficient to protect human health and the environment.

EPA is requiring landfill operators to make random inspections of incoming waste or to take other steps to ensure that incoming loads do not contain regulated hazardous wastes. Therefore, landfills may require sludge disposers to prove that their sludge is not hazardous. Domestic wastewater sludges are rarely ignitable, corrosive, or reactive. However, some sludges may be toxic and are therefore classified as hazardous wastes. Many landfills are already requiring that sludges pass the *Toxicity Characteristics Leaching Procedure* (TCLP) analysis before disposal. In March 1990, EPA mandated use of the TCLP test to determine if a solid waste is toxic. This rule applies to co-disposed wastewater plant sludges as well.

EPA now prohibits the disposal of bulk or non-containerized liquid wastes in landfills. Subtitle D requires that the owner or operator of the landfill use the *Paint Filter Liquids Test* (PFLT) to determine whether the wastes (including septic wastes and municipal wastewater sludges) are liquid. This simple test is performed by placing a representative sample of the sludge in a mesh No. 60 paint filter (available at paint stores), then allowing it to drain for five minutes. The sludge is considered a liquid waste if any liquid passes through the filter during the five-minute period. Dewatering sludges to approximately 18 to 20 percent dry solids will usually be adequate for removing the free liquid. Most sludge disposers already dewater their sludge because landfills generally do not accept liquid wastes.

(11) *Recordkeeping*—Any groundwater monitoring data, gas monitoring data, inspection reports, and training procedures must be retained by the facility.

4.3.3 CLOSURE CRITERIA

The regulations cover the performance standard of the completed municipal solid waste landfills to minimize the post-closure formation and release of leachate and explosive gases to the air, groundwater, or surface water.

(1) *Closure Plan*—The owner or operator must prepare a closure plan which specifies the following:
(a) Overall description of closure, (b) estimation of the maximum extent of operation that will open at any given time during the life of the landfill, (c) estimation of the maximum inventory of waste on site, and (d) description of the final cover system. Many of these issues have been discussed in Chapter 3.

(2) *Timings*—The closure must begin within 30 days of the final receipt of waste. The owner or operator must submit a third party certification that the units are closed in accordance with the closure plan.

(3) *Post-Closure Care Requirements*—There are two phases of post-closure care: (a) Phase one consists of 30 years with the requirements of final cover maintenance, leachate collection system operation and maintenance, groundwater monitoring, and gas monitoring, and (b) Phase two consists of groundwater and gas monitoring. The length of phase two is determined by the governing state agency.

(4) *Documentation*—The owner or operator must prepare a written post closure plan which includes the following: (a) a description of the monitoring and maintenance of each unit and the frequency of such monitoring, (b) a contact person for both phases of post-closure, and (c) a description of the planned uses of the property. The owner or the operator must make a notation in the deed of the property that the

property was a municipal solid waste landfill. The owner or operator must submit a third party certification of post-closure care for both phases.

(5) *Financial Assurance*—A statement of the financial instruments available to the owner or operator must be made to maintain the necessary funding for closure and post closure care. The requirements of this section apply to all municipal solid waste landfills except those owned by a state or federal government entity. The estimates for closure and post closure care shall be based on a third party performing the work. The estimate is based also on closing the facility at the point when such closure would be the most expensive. The estimate is adjusted annually during the active life of the facility and also if there are any changes to the closure or post-closure plans.

4.3.4 DESIGN CRITERIA

The regulations provide details of design criteria that must be met for MSWLFs:

(1) *New MSWLF Units and Expansions*—The MSWLFs must be installed with a composite liner. The designed and constructed leachate system shall maintain less than 30-cm depth of the leachate over the liner. The design must ensure that the concentrations of 24 chemicals will not be exceeded in the uppermost aquifer at the relevant point of compliance specified by the state. The maximum contaminant levels of 24 chemicals are also given in Table 4.1.

(2) *Composite Liner*—All MSWLFs shall have a composite liner system, leachate collection system and cover system, for containing the leachate, to ensure a very low groundwater carcinogenic risk level.

A composite liner system is constructed of two components. The upper component must consist of a minimum of 0.7 mm (30-mil) flexible membrane liner (FML), and the lower component of at least a 0.6 m layer (2 ft) of compacted soil with a hydraulic conductivity of no more than 1×10^{-7} cm/sec. The FML components consisting of high density polyethylene (HDPE) shall be at least 1.5 mm (60-mil) thick, and must be installed in direct and uniform contact with the compacted soil component. The liner installation details are shown in Figures 3.6–3.8, and discussed in Section 3.8.1.2.

When considering the design to comply with low carcinogenic risk the following must be considered: (a) hydrogeology, (b) climate, (c) volume and physical characteristics of the leachate, (d) proximity of groundwater uses, (e) the quality, quantity and direction of groundwater flow, (f) availability of alternative drinking water supplies, and (g) practicable capability of the owner or operator.

Part 258 requires final cover for all landfills that receive waste on or after October 9, 1991. However, existing active landfills are not required to remove disposed waste in order to install liners or leachate collection systems. If an existing landfill cannot meet the location restrictions, it will have to close by October 9, 1996. The five-year period will allow users of the landfill to find alternative facilities or methods of disposal.

4.3.5 GROUNDWATER-MONITORING AND CORRECTIVE ACTION

The groundwater monitoring requirements can be waived only if the owner or operator can demonstrate that there is no potential for migration of hazardous constituents from the unit to the uppermost aquifer during the active life, closure period, and post closure period.

Within six months of the effective date of the rules, the State must specify a schedule for owners or operators to comply with the requirements of groundwater monitoring and corrective action. All MSWLFs must comply

TABLE 4.1. Maximum Contaminant Levels in Uppermost Aquifer.

Chemical	MCL (mg/L)
Arsenic	0.05
Barium	1.0
Benzene	0.005
Cadmium	0.01
Carbon tetrachloride	0.005
Chromium (hexavalent)	0.05
2,4-Dichlorophenoxy acetic acid	0.1
1,4-Dichlorobenzene	0.075
1,2-Dichloroethane	0.005
1,1-Dichloroethylene	0.007
Endrin	0.0002
Fluoride	4
Lindane	0.004
Lead	0.05
Mercury	0.002
Methoxychlor	0.1
Nitrate	10
Selenium	0.01
Silver	0.05
Toxaphene	0.005
1,1,1-Trichloromethane	0.2
Trichloroethylene	0.005
2,4,5-Trichlorophenoxy acetic acid	0.01
Vinyl Chloride	0.002

Source: U.S. EPA (1991).

within five years of the effective date. The State must establish a well defined compliance schedule pertaining to notifications, monitoring program, sampling, and analysis of chemical constituents specified in Appendix I and II,** corrective measures, and placement of records. Appendix I and II are given in Subtitle D (U.S. EPA 1991).

(1) *Groundwater Monitoring Systems* – Groundwater monitoring requires installation of groundwater monitoring wells. (a) Wells must be as close as possible to the site boundary. There must be sufficient wells at appropriate locations and depths to obtain samples from the uppermost aquifer that represent unimpacted groundwater quality, and represent downgradient quality passing the waste unit, (b) Wells must be cased to insure integrity of the borehole, (c) All well installations, construction, and abandonment must be documented, (d) All wells must be operated and maintained through the active life of the facility, as well as through closure and post closure care, (e) The number, spacing, and depths of the monitoring wells must be based on site specific criteria approved by the state, and (f) The owner or operator must provide a site characterization of aquifer thickness, aquifer flow rate, aquifer flow direction, geologic units overlying the uppermost aquifer, thickness, stratigraphy, lithology, hydraulic conductivities, porosities, etc.

(2) *Groundwater Trigger Level* – The state must establish trigger levels protective of human health and the environment, for all Appendix II constituents (Appendix II is given in Subtitle D). These levels must be based on *Maximum Contaminant Levels* (MCLs) established under the *Safe Drinking Water Act,* health based levels for constituents not having MCLs, and health risks for carcinogens. If a background level in the water quality exceeds one of these levels then the background concentration becomes the trigger level.

(3) *Sampling and Analysis Requirements* – The operating record must include a sampling and analysis plan which details sample collection, sample preservation and shipment, analytical procedures, and chain of custody control. Groundwater levels must be measured to determine the direction and rate of flow. Background water quality must also be determined on hydraulically upgradient wells. Finally, a statistical analysis of the data is required.

**Appendix I in Subtitle D contains 62 constituents for detection, and Chemical Abstract Service Registry Number (CASRN). Appendix II in Subtitle D contains common names, CASRN, suggested EPA analytical methods, and Practical Quantitation Limits (PQLs) for 213 inorganic and organic chemicals used in government regulations.

The monitoring program for groundwater is broken down into Phase I monitoring, Phase II monitoring, and Corrective Action. The Phase I monitoring program calls for at least the semi-annual sampling of 24 inorganic parameters and certain volatile organic compounds. If there is a statistically significant increase over background of the Phase I parameters, the state must be notified within 14 days, and Phase II monitoring is established. Phase II monitoring consists of 213 parameters and must be implemented within 90 days of the statistical trigger. Phase II monitoring is continued at a frequency specified by the state, and, if contamination is proved, a corrective action program is developed. The content of this program is selected by the State.

4.3.6 CLOSURE AND POST-CLOSURE CARE

(1) *Closure Criteria*—The owner or operator of all MSWLF units must install a final cover system that is designed to minimize infiltration and erosion. The final cover system must comprise an erosion layer underlain by an infiltration layer as follows:

> The infiltration layer must comprise a minimum of 45.7 cm (18 inches) of earthen material that is capable of sustaining native plant growth. The state may approve an alternative equivalent final cover design. The owner or operator must prepare a written closure plan defining final cover, estimate maximum inventory of wastes on-site over the active life of the MSWLF, notify state about the closure plan, and follow notification and action guidelines.

(2) *Post-Closure Care Requirements*—Following closure of each MSWLF unit, the owner or operator must conduct post-closure care. Post closure care must be conducted for 30 years except as approved otherwise by the State. The post closure care requirements consist of maintaining the integrity and effectiveness of the final cover, operating the leachate-collection system, monitoring groundwater, and maintaining and operating the gas monitoring system. The owner or operator must also provide a description of planned uses of the property, as well as name, address, and telephone number of the person or office to be contacted about the facility during the post-closure period. All notification and post-closure action plans must be followed.

4.3.7 FINANCIAL ASSURANCE

The owner or operator of the MSWLF must have a detailed written estimate of the cost of hiring a third party to perform closure and post-closure care over the specified period. During the active life of the MSWLF, these estimates must be revised annually and financial assurance be increased accordingly. The financial assurance must include post-closure care and corrective actions. The allowable financial assurance mechanisms are

- Trust fund
- Surety bond guaranteeing payment or performance
- Letter of credit
- Insurance
- Corporate-financial guarantee
- Local government guarantee
- State-assumption of responsibility
- Multiple financial mechanisms

Each of these mechanisms are fully described in Subtitle D.

4.4 EFFECTS OF SUBTITLE D ON MUNICIPAL SOLID WASTE LANDFILLS

Under *Subtitle D* of the *Resource Conservation and Recovery Act* (RCRA), the states are required to adopt and enforce regulations that encompass location restrictions, operating requirements, design standards, groundwater monitoring and corrective action requirements, closure and post-closure care requirements, and financial assurance standards. The regulation is self-implementing, meaning that the landfill owners and operators are required to be in compliance regardless of the state regulations. The regulations took effect on October 9, 1993.

Many changes are expected in MSWLF practices. Many landfills may have to close due to the financial burden resulting from these regulations, while others will improve the design and operation for environmental protection. It is also expected that many products may appear in the market that may provide improved design, save valuable space, trim costs, and provide greater protection to the environment.

4.4.1 LANDFILL CLOSURE

If an existing landfill cannot meet the regulatory constraints, it may close. EPA did not estimate the number of landfills that will have to close as a result of new regulations. However, the number of landfills and available landfill space has been shrinking in recent years and the trend is expected to continue.

Repa and Sheets (1992) reported the results of a survey of many sources, conducted on landfill practices in North America. Although there are inconsistencies in these estimates, a reasonable estimate is 6600 MSWLFs in existence today. The best estimates on landfill closings, openings, and expansions since 1986–1991 are 2,216, 364, and 407 respectively. These data clearly show an average of (1) 63 landfill closures per state, (2) six new landfill permits per state, (3) 10 landfill expansion permits per state, and (4) approximately 10 closures for every new landfill permitted. The survey data tend to support the presumption that new landfills generally have greater capacities and may be providing a net increase in capacity over the closure of several smaller facilities. Public opposition to landfill siting is perhaps the biggest obstacle to increasing capacity. Repa and Sheets (1992) also reported that EPA projects declining landfill numbers in the future. By the end of this century, over 2000 additional landfills will be closed, and 28 states in the United States will need to replace all of their current landfill capacity, or will need alternative solid waste management methods. Currently, the average municipal solid waste generation is over 200 million tons per year. EPA estimated that 64 per cent of the solid waste is disposed of in landfills, 18 percent is recycled and 18 percent is incinerated. By the year 2000, municipal solid waste generation rate is expected to increase. Even with increased recycling and incineration, approximately 49 percent of total MSW will need to be landfilled. This means that 82 percent of the current landfill capacity will still be needed by the end of the century (Repa and Sheets 1992).

4.4.2 COMPLIANCE

The new RCRA subtitle D regulations will apply to all MSWLFs except those owned and operated by state or the federal government. During the winter of 1990–1991, the *National Solid Waste Management Association* (NSWMA) surveyed its members and other selected facilities, to collect data on physical and operational characteristics, and on design and environmental protection features of municipal landfills. In this survey, 87 percent of the facilities were comprised of privately operated landfills. The country was divided into seven geographical regions: Northeast, Mid-Atlantic, South, Midwest, West Central, South Central, and West.

The results of the survey as reported by Repa and Sheets (1992) are summarized in Table 4.2. It may be noted that most of the landfill owners and operators have initiated the necessary steps needed to meet the requirements of Subtitle D. A number of these landfills are nearly in compliance with RCRA Subtitle D or exceed these requirements. As an example, the world's largest landfill site, the *Fresh Kills,* located on Staten Island, is utilizing over 220 ground monitoring wells (Anonymous, 1992). In addition to their efforts which improve the operation and design of landfills, many landfill owners support local recycling or provide other services. The NSWMA survey also indicated that 52% had residential refuse drop-off

TABLE 4.2. Summary Results of NSWMA Survey of Municipal Solid Waste Landfills in USA.

PHYSICAL/OPERATIONAL CHARACTERISTICS
98% exclude hazardous wastes
99% cover solid waste daily
65% monitor gases
66% control gas
   51% flare
   21% recovery
96% control access
87% control surface water run-on/run-off
65% monitor surface water quality
78% monitor for heavy metals
68% monitor for organics

DESIGN
Liners
   81% utilize some type of liner
      21% natural
      51% clay
      13% composite[a]
      3% composite and FML[b]
      6% double composite
      3% single FML
      3% double FML

Leachate Collection and Treatment
   68% utilize leachate collection
      15% use recirculation for disposal
      17% use on-site treatment
      62% use off-site treatment
      6% use combination

Groundwater Monitoring
   95% utilize groundwater monitoring
      Number of wells per facility:
         Maximum 357
         Average 19
         Minimum 1
      Parameters Monitored
         95% general water quality parameters (e.g. pH, TDS, chloride)
         88% heavy metals (e.g., lead, cadmium)
         82% organics (e.g., volatile organic compounds)
      Cover
         100% use top soil cover
         12% soil cover only
         71% soil/clay

CLOSURE AND POST-CLOSURE CARE
93% prepared closure plan
83% filed closure financial assurance
91% prepared post-closure plan
100% filed post-closure care financial assurance

[a]Clay and flexible membrane liner.
[b]Clay and flexible membrane liner.
Source: Repa and Sheets (1992).

TABLE 4.3. Comparison of NSWMA and EPA survey data.

Control Measures	Percent of Total	
	NSWMA	EPA
Surface water control	87	61
Surface water monitoring	65	15
Methane monitoring	66	7
Methane controls	66	2
Liners (on-site and engineered)	81	28
Leachate collection systems	68	11
Groundwater monitoring	95	36

Source: Repa and Sheets (1992).

on-site, 37% had recycling drop-offs, 29% had materials recovery facilities, and 21% had compost facilities.

The U.S. Environmental Protection Agency also conducted a similar survey in 1988. Although this survey was a few years earlier than the NSWMA survey, similar results were developed by EPA. Repa and Sheets (1992) provided a comparison of the findings of both surveys. A general comparison, as reported by Repa and Sheets (1992) is provided in Table 4.3.

4.4.3 IMPROVED LINERS AND OTHER PRODUCTS

New requirements under Subtitle D, although considered a financial burden for many landfills, have resulted in several innovative and improved products being developed and marketed which are easy to install, save space and reduce cost. Among these are liners, edge welders, daily cover, gas and leachate collection and monitoring systems, and on-site leachate treatment plants.

Sanitary landfill liners today have much higher levels of performance than in the past. They experience fewer problems with cracking, are easy to install, have stronger and quicker edge welding, and have more conformity to irregularities that may exist at the base of a landfill. Woods (1992) reported many types of innovations in liners, including liners of more durable material, wider sheets to reduce joint line, and liners made of three-dimensional nylon matting or pulverized rubber material that can withstand pressures of 550 kPa (80 psi)† or equivalent of 50 m (164 ft) head of water. These liners may be up to 2 cm (0.75 inch) thick, and can

†kPa = kN/m^2, 1 lb/in^2 = 6.89 kN/m^2.

conveniently eliminate the need for 0.6 m (2 ft) of compacted soil layer, thus providing additional volume for solid waste. Most of these liners assist in providing better leachate flow to the collection point. Other laminated composite materials help to filter and drain leachate into a collection system. Still others may allow leachate collection capability on steep slopes, where a conventional aggregate would be unstable.

Several types of reusable cover panels for daily cover use are available. These panels take only a few minutes to install at the end of the day, and to remove the next day for solid waste compaction. Such products save 0.15 m (6 inch) of landfill thickness each day resulting in more space for solid wastes. Likewise, synthetic foam products, sprayed at the end of the day, replace the need for daily compacted soil cover. These foam products, 2–12 cm (1–5 inch) thick in place (which provide equivalent protection of 0.15 m (6 inches) of compacted daily soil cover), are crushed to only a fraction of their depth, making room for more solid wastes (Woods 1992).

4.4.4 COST INCREASE OF LANDFILLING

The cost of implementing Subtitle D requirements to MSWLFs will vary greatly, depending upon to what extent regulations were previously imposed upon site selection, design, operation, and closure. However, based on different scenarios, the estimated increases range from $1 to $24 per wet ton. The average increase based on EPA estimates for landfill disposal or tipping fees is $2 per wet ton (Black and Veatch 1992). EPA officials predict that the cost of implementing the new rules will run about $330 million per year (Woods 1992).

4.5 REFERENCES

Anonymous. 1992. "How Largest Landfill Samples Groundwater," *Water, Engineering and Management*, 139(10):29.

Black and Veatch, Inc. 1992. *Sludge Regulations Update*, Issue I, pp. 1–6.

Pfeffer, J. T. 1992. *Solid Waste Management Engineering*, Englewood Cliffs, NJ: Prentice Hall.

Repa, E. W. 1992. "Subtitle D: Update: EPA Promulgates Long-Awaited Landfill Rules," *Waste Age*, 22(10):69–72, 156.

Repa, E. W. and S. K. Sheets. 1992. "Subtitle D's Effects on Private Industry," *Waste Age*, 23(4):267–272.

U.S. Congress. 1965. *Solid Waste Disposal Act*, Title II of (PL 89-272), 89th U.S. Congress.

U.S. Congress. 1970. *The Resource Recovery Act*. (PL91-512), 91st U.S. Congress.

U.S. Congress. 1972. *The Marine Protection, Research, and Sanctuaries Act of 1972*, (PL 92-532), 92nd U.S. Congress.

U.S. Congress. 1976a. *The Resource Conservation and Recovery Act,* (PL 89-272), 89th U.S. Congress.

U.S. Congress. 1976b. *Toxic Substances Control Act,* (PL 94-469), 94th U.S. Congress.

U.S. Congress. 1980. *The Comprehensive Environmental Response Compensation and Liabilities Act of 1980, (CERLA),* (PL 96-510), 96th U.S. Congress.

U.S. Congress. 1984. *The Hazardous and Solid Waste Amendments of 1984,* (PL 98-616), 98th U.S. Congress.

U.S. Congress. 1986. *The Superfund Amendments and Reauthorization Act of 1986,* (PL 99-499), 99th U.S. Congress.

U.S. Environmental Protection Agency. 1991. *Solid Waste Disposal Facility Criteria; Final Rule,* 40 CFR Part II, 257 and 258, Subtitle D of the Resource Conservation and Recovery Act (RCRA).

U.S. Environmental Protection Agency. 1993. *Standards for the Use or Disposal of Sewage Sludge,* 40 CFR 503, February 19.

Woods, R. 1992. "Building a Better Liner System," *Waste Age,* 23(3):26–32.

CHAPTER 5

Moisture Routing and Leachate Generation

5.1 INTRODUCTION

To plan and design a sanitary landfill effectively, it is important to understand what takes place within a landfill after operations have been completed. Solid wastes undergo a number of simultaneous biological, physical, and chemical changes. The water moves through the fill and carries with it extractable chemicals. The decomposition, stabilization, and extraction of pollutants from a landfill depend upon several factors: composition of the wastes, degree of compaction, amount of moisture present, presence of inhibiting materials, rate of water movement, and temperature. In this section the decomposition process in a sanitary landfill and leachate generation mechanism are presented.

5.2 SOLID WASTE DECOMPOSITION IN A SANITARY LANDFILL

The production of leachate from municipal sanitary landfills is an important environmental problem. Many factors interact to produce variable quantity and quality of leachate from landfills. Leckie et al. (1979) reported some of the most relevent factors that influence leachate generation. These factors are: annual precipitation, runoff, infiltration, evaporation, transpiration, freezing, mean ambient temperature, waste composition, waste density, initial moisture content, and depth of the landfill. These authors also reported that the stabilization of solid waste placed in a sanitary landfill, and the quality of leachate are principally the result of physical, chemical, and biological processes; thus, additional variables are water movement, macronutrients, micronutrients, and the presence or absence of toxic or inhibitory elements and compounds.

As a sanitary landfill ages under a constant water input, it passes through a succession of stages. Little or no leachate is produced until the landfill reaches its field capacity (becomes saturated). Because of compaction and squeezing, the leachate produced prior to saturation will, in general, depend on the quantity of water initially present.

The decomposition process in a sanitary landfill has been discussed by numerous researchers. Qasim (1965), Qasim and Burchinal (1970a), Brunner and Keller (1972), Pfeffer (1992) and Tchobanoglous et al. (1993) reported that the decomposition process primarily occurs in three stages. Initially, aerobic decomposition predominates. This phase is generally very short because of the limited amount of oxygen in the landfill and the high biochemical oxygen demand (BOD) of the solid waste. During this phase a large amount of heat is produced, raising the landfill temperature well above the ambient temperature. Leachate produced during this phase would be expected to dissolve highly soluble salts, such as NaCl and others.

As oxygen is depleted, decomposition caused by faculative anaerobic organisms generally predominates. During this stage of anaerobic decomposition, large amounts of volatile fatty acids such as acetic acid, and carbon dioxide, are produced. These acids reduce the pH to between 4 and 5. The low pH helps to solubilize inorganic materials, which, along with the high concentrations of volatile acids, produce a high ionic strength. The high volatile acids concentrations also contribute to the high strength chemical oxygen demand (COD) often found during this phase. The redox potential is reduced to below zero mv.

The second stage of anaerobic decomposition occurs when the population of methane producing bacteria builds up (Gaudy and Gaudy 1988). Methane bacteria are strict anaerobes and require neutral pH (6.6 to 7.3). Volatile acids produced by faculative anaerobes, and other organic matter are converted to methane and carbon dioxide. Thus, the volatile acid concentration is reduced to lower levels, and the gas composition becomes a mixture of carbon dioxide and methane. The pH starts to rise which encourages methane production. At near neutral pH, fewer inorganic materials are solubilized and conductivity falls. However, some materials continue to solubilize as the decomposition process continues. The redox potential should be lower than that during the early stages of anaerobic processes, reflecting the potential for methane production and increase in pH.

Eventually, as a landfill ages, the rate of bacterial decomposition may decrease due to substrate depletion. Slowly, portions of the landfill may reestablish aerobic conditions as oxygenated water continues to percolate into the fill. Tchobanoglous et al. (1993), and Peavy et al. (1985), provided generalized equations for aerobic and anaerobic decomposition of organic matter. These equations are discussed in Chapter 2 [Equations (2.6) and (2.7)].

Stanforth et al. (1979) developed generalized degradation curves expressing pH, oxygen, carbon dioxide, methane, acetic acid, solubilized salts, and redox potential. These generalized curves are illustrated in Figure 5.1. The decomposition in a sanitary landfill can continue for many years, as long as some organic material is available for bacterial activity. The rate depends on many factors including water movement, pH, temperature, degree of compaction, age of fill, and composition of solid wastes. As degradation occurs, the volume of the original solid waste is reduced, in effect allowing greater penetration by rain in some cases. In particular, decomposition under relatively dry conditions stops and materials can remain unaltered for decades.

Besides chemical contaminants in liquid state, major gaseous production of landfill decomposition are eventually emitted to the atmosphere. Woodwell (1984) reported that these gases were previously considered

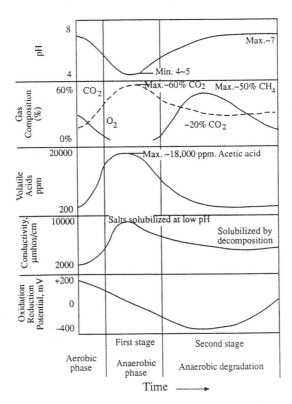

Figure 5.1 Generalized degradation curve of a theoretical landfill during decomposition process. Source: Adapted from Stanforth, Ham, and Anderson (1979), with permission from the Water Environment Federation.

relatively benign (except for the explosive threat of methane), but they are now considered as prime factors in the global warming phenomenon. In recent years, efforts have been made to recover methane from sanitary landfills to use it as an energy resource (Flanagan 1988; Vasuki 1988; U.S. Congress 1989). This subject has been covered in Chapter 3.

5.3 LEACHATE GENERATION FROM SANITARY LANDFILLS

Water passing through a sanitary landfill carries with it various dissolved and suspended materials. Generally, as more water flows through the solid wastes, more pollutants are leached. It is therefore important to review the methods that can be used to estimate the amount of leachate generation at a sanitary landfill site. Remson et al. (1968), Fenn et al. (1975), Dass et al. (1977), Lu et al. (1984), Korfiatis and Demetracopoulos (1984), Pfeffer 1992, and Tchobanoglous et al. (1993), have suggested techniques that utilize the water budget analysis through a landfill.

5.3.1 WATER BUDGET

Fenn et al. (1975) and Dass et al. (1977), presented a water budget analysis. The principle source of moisture is precipitation over the landfill site. A part of this moisture results in surface runoff, a part is returned to the atmosphere in the form of evapotranspiration from the soil and plants' surfaces, and the remainder adds to the soil moisture storage. Whenever moisture exceeds the field capacity of the soil it percolates down into the solid waste. Field capacity is the maximum moisture that can be retained without continuous downward percolation by gravity. The addition of moisture to solid waste over a period of time saturates the solid waste to its field capacity. At that stage the moisture from the solid waste emerges as leachate. The rate of moisture percolation to the solid waste after the initial delay is equal to the rate of leachate generation. Various components of the moisture used in the water budget are shown in Figure 5.2. Important factors that govern the rate of percolation are precipitation, surface runoff, evapotranspiration, and soil moisture storage. These factors are briefly described below. Additional information may be found in works published by Fenn et al. (1975), Dass et al. (1977), Lu et al. (1981), and Lu et al. (1985).

5.3.1.1 Precipitation

Precipitation includes all water that falls from the atmosphere to the area under consideration. It may occur in a variety of forms including

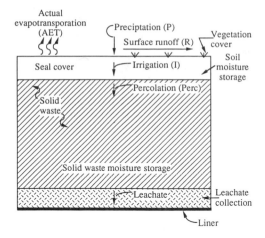

Figure 5.2 Moisture components at a sanitary landfill. Source: Fenn et al. (1975).

rainfall, snow, hail and sleet. Once precipitation strikes the ground, it will produce surface runoff, evaporation and percolation.

Precipitation varies geographically and seasonally. The amounts of precipitation sometimes vary considerably within a short distance. Therefore, reliable precipitation data for the area under consideration must be developed. Important data sources on precipitation are publications of the National Weather Service, and various other federal and state agencies that keep historical precipitation data.

5.3.1.2 Surface Runoff and Infiltration

A portion of incident precipitation runs off the site and is lost to the overland flow before it has as chance to infiltrate. The amount of surface runoff depends upon factors such as intensity and duration of the storm, the surface slope, the permeability of the soil cover, and the amount and type of vegetation. Several methods have been proposed to estimate the runoff from, or percolation into a sanitary landfill. These methods include field measurements and empirical relationships. The most commonly used relationship is the rational method. Empirical *runoff coefficients,* such as those used for designing surface water drainage systems, must be used with caution.

Since landfill sites have potholes and cave-ins, the surface runoff does not occur freely like it does over natural ground cover. Therefore, the use of these runoff coefficients may yield unrealistically high runoff values for landfills, and may grossly underestimate the amount of percolation. Considerable work has been done in developing runoff coefficients for application in various agricultural and engineering situations (Jens et al. 1949; Joint Committee of WPCF and ASCE 1970; U.S. Soil Conservation Service 1972; Lutton et al. 1979). Empirical values for obtaining runoff coefficients for different topography, soil, and cover are given in Table 5.1 (Lutton et al. 1979). The runoff coefficients used in the rational method for estimating quantities of storm runoffs are provided in Table 5.2 (Joint Committee of WPCF and ASCE 1970). In most cases, it is expected that the surface runoff coefficients for sanitary landfill conditions may lie within the range of 0.07 to 0.2.

Another method, called the *curve number method,* has been proposed by the U.S. Soil Conservation Service to predict surface runoff from agricultural lands (U.S. Soil Conservation Service 1975). The curve number method uses an empirical equation from which the runoff is calculated. In addition to rainfall, soil type, and land cover, the method accounts for land use and antecedent moisture condition. Lutton et al. (1979) and Bagchi (1990) provided examples for calculating runoff by this method.

In colder regions, the infiltration from snow depends upon the ground condition, temperature and rate of snow melt. Two methods have been proposed to estimate the runoff from snow covered areas. These are the *degree day* method suggested by the U.S. Soil Conservation Service (1975), and the *U.S. Army Corps of Engineers,* method (Lu et al. 1981, 1985; Chow et al. 1988). Bagchi (1990) provided an example of runoff calculation from the degree day method.

TABLE 5.1. Empirical Values for Obtaining Runoff Coefficient.

Site Description	Value, v'
Topography	
Flat land, with average slopes of 0.02–0.06%	0.30
Rolling land, with average slopes of 0.3–0.4%	0.20
Hilly land, with average slopes of 2.8–4.7%	0.10
Soil	
Tight impervious clay	0.10
Medium combinations of clay and loam	0.20
Open sandy loam	0.40
Cover	
Cultivated lands	0.10
Woodland	0.20

Add values v' for topography, soil, and cover, and subtract from unity to obtain runoff coefficient.
Source: After Lutton et al. (1979).

TABLE 5.2. Typical Values of Coefficient of Runoff.

Surface Type	Coefficient of Runoff
Bituminous Streets	0.70–0.95
Concrete Streets	0.80–0.95
Driveways, Walks	0.75–0.85
Roofs	0.75–0.95
Lawns; Sandy Soil	
Flat, 2%	0.05–0.10
Average, 2–7%	0.10–0.15
Steep, 7%	0.15–0.20
Lawns, Heavy Soil	
Flat, 2%	0.13–0.17
Average, 2–7%	0.18–0.22
Steep, 7%	0.25–0.35

Source: Adapted from Joint Committee of WPCF and ASCE (1970), with permission from Water Environment Federation.

5.3.1.3 Evapotranspiration

The amount of moisture available for evapotranspiration at a landfill site is affected by the type of soil and vegetation. A desirable feature of sanitary landfill design is to increase the evapotranspiration in an effort to reduce the leachate production (Dass et al. 1977). Evapotranspiration at a site is either estimated or measured. Actual measurement of evapotranspiration is a more accurate prediction of real losses. Many proposed estimation methods utilize lysimeter measurement or pan evaporation reduced by the pan evaporation coefficient. These methods are given in most surface water hydrology books and technical papers (Jens et al. 1949; U.S. Soil Conservation Service 1972; Viessman et al. 1977; Linsley and Franzini 1979; Chow et al. 1988).

Many empirical evaporation equations have been proposed by several researchers. Veihmeyer (1964) provided a discussion on methods and equations suggested by Hedke, Lowry-Johnson, Blaney-Criddle, Blaney-Morin, Penman, Hargreaves, and Thornthwaite. Among all these methods and equations, the procedures provided by Thornthwaite and Mather (1955, 1957) are the most commonly applied. This method utilizes tabulated values of (a) heat index, (b) temperature, and (c) correction factor. Additional information is provided later in this section. The consumptive use of crops or vegetation provides a gauge for the prediction of the amount of irrigation needed to supplement the natural precipitation in arid areas. Table 5.3 provides some example values of consumptive use from

TABLE 5.3. **Examples of Consumptive Use of Water.**

Crop	Consumptive Use (m/year)
Alfalfa	3.5
Pasture	3.5
Wild Hay	2.6
Grass/Weeds	1.8
Small Grain	1.6
Oats	1.2
Wheat	1.3

Source: After Lutton et al. (1979).

the literature (Lutton et al. 1979). The calculation procedure for evapo-transpiration is provided in the example for water balance calculations. (Thornthwaite and Mather 1955, 1957).

5.3.1.4 Soil Moisture Storage

The moisture content in the soil is continually changing: increasing due to infiltration and decreasing due to evapotranspiration. The depletion of

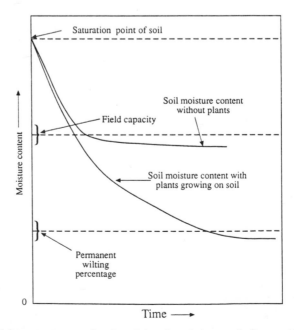

Figure 5.3 Moisture content as a function of time for a drainage soil. Source: After Remson et al. (1968), with permission from The Journal of Environmental Engineering Division, ASCE.

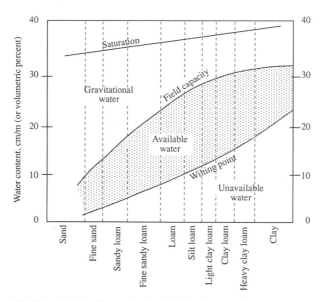

Figure 5.4 Water-holding characteristics of soils. Source: After Lutton et al. (1979).

moisture due to evapotranspiration is limited to an upper soil zone defined by the *effective root zone depth*. In the water balance method, it is important to account for the change in the moisture storage within the soil cover. The maximum moisture that a soil can hold against the pull of gravitational force is the *field capacity*. The minimum moisture a soil loses due to vegetation is its *moisture content at wilting point*. The difference between the two limits is the *moisture holding capacity* available to the plants from the soil. The higher the available moisture holding capacity of the soil, the more moisture it may lose to the atmosphere by evapotranspiration. Figure 5.3 shows the drainage properties of an unsaturated permeable soil. Initially, the soil drains rapidly until the field capacity is reached (Remson et al. 1968). At field capacity, the soil remains essentially at that moisture content unless it loses moisture in other ways. The soil within the root zone can be dried by evapotranspiration to moisture contents considerably below the field capacity. At the *permanent wilting point,* the remaining moisture is essentially unavailable for withdrawal by plants. Figure 5.4 shows the water holding characteristics of several soils (Lutton et al. 1979).

To illustrate the use of Figure 5.4, consider a 0.6 m top soil cover of silty loam over a sanitary landfill.

The field capacity $= 28$ cm/m $\times$ 0.6 m $= 16.8$ cm
The wilting point $= 10.5$ cm/m $\times$ 0.6 m $= 6.3$ cm
The storage capacity $= (16.8$ cm $- 6.3$ cm$) = 10.5$ cm

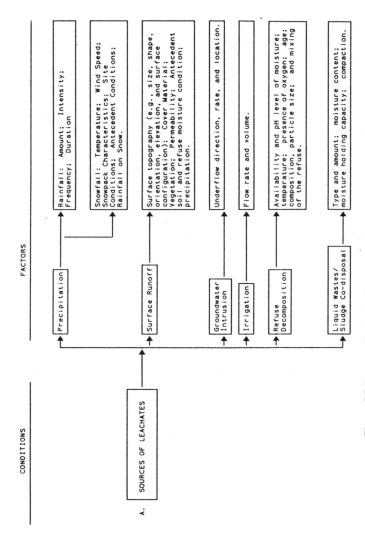

Figure 5.5 Factors affecting leachate generation. Source: Lu et al. (1981).

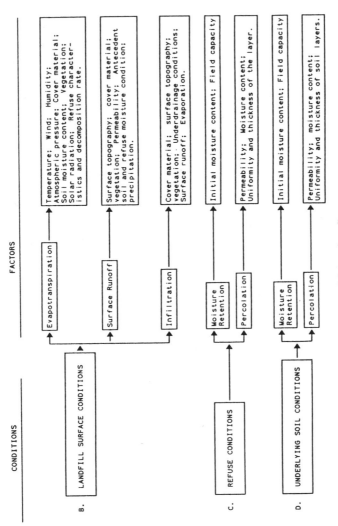

Figure 5.5 (continued).

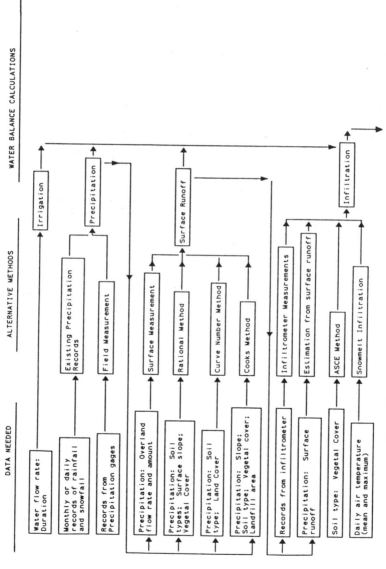

Figure 5.6 Flow chart of water balance calculation. Source: Lu et al. (1981).

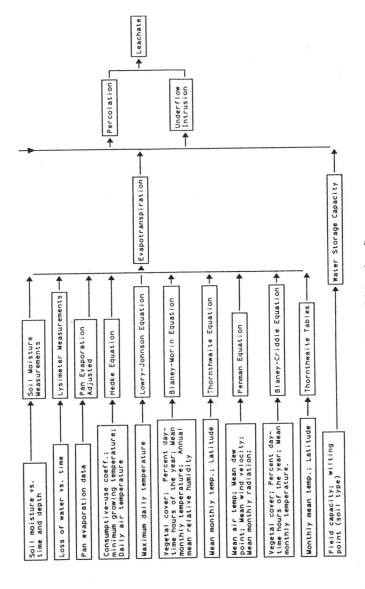

Figure 5.6 (continued).

111

5.3.2 MOISTURE ROUTING THROUGH SANITARY LANDFILLS

Water balance calculations for sanitary landfills can be conducted by several methods. Lu et al. (1981) summarized the major factors which affect the conditions and, ultimately, leachate generation. These factors are provided in Figure 5.5. Lu et al. (1981, 1984) summarized the techniques available for leachate volume estimation. These techniques are itemized in Figure 5.6. As shown in this figure, a combination of 240 different methods are available for leachate generation calculations. Data needed for each method is listed in the first column of Figure 5.5 (Lu et al. 1981, 1984).

Fenn et al. (1975) provided a detailed method for conducting water balance calculations. Dass et al. (1977) showed moisture balance calculations for a sanitary landfill in Wisconsin. Pfeffer (1992) gave a simplified moisture routing based on net percolation and net water loss. Tchobanoglous et al. (1993) provided an example for leachate quantity calculation based on water budget analysis. Because of the tedious procedure involved in the calculation of leachate generation factors and moisture routing data, attempts have been made to develop computer simulation models for estimation of leachate generation. Lutton et al. (1979) adapted the U.S. Department of Agriculture Hydrograph Laboratory (USDAHL) model and applied this model to actual landfill conditions. A more complete model, designed specifically for landfill sites, is the HSSWDS (Hydrologic Simulation on Solid Waste Disposal Sites) model, developed by Perrier and Gibson (1980). This model was intended to assist the owners, operators and designers of the solid waste disposal sites to develop design and operation procedures. Another computer program, the Hydrologic Evaluation of Landfill Performance (HELP) developed by the U.S. Army Corps of Engineers, uses a quasi-two dimensional hydrologic model for movement of water across, into, through, and out of a landfill (Schroeder et al. 1984). Garner and Conrad (1986), reported the application of this model for the New York City Department of Sanitation. Several landfills were evaluated using this model for different design conditions. Peyton and Schroeder (1988) provided long-term simulations of 17 landfill cells using the HELP computer model. These landfills were located in different regions of the country. The results indicated that the HELP model can be a very useful tool for designing and evaluating landfills.

U.S. EPA (1992) developed a Multimedia Exposure Assessment Model (MULTIMED) that simulates the transportation of contaminants released from the waste disposal facility into the multimedia environment. The model simulates the releases into air or soil, including unsaturated and saturated zones, and possible interception of the surface contaminant plume by surface streams. Analytical and semi-analytical solution techniques are utilized to solve the mathematical equations describing flow and transport.

Two methods are presented here for estimating the leachate generation from sanitary landfills. These are: the procedure reported by Fenn et al. (1975) and the simplified routing method by Pfeffer (1992) and Tchobanoglous et al. (1993).

5.3.3 WATER BALANCE CALCULATIONS

The water routing through a sanitary landfill basically consists of two phases: routing through the soil cover, and routing through the compacted solid waste beneath (Figure 5.2). The soil cover directly interacts with the atmosphere, and will determine the amount of infiltration into the soil and percolation into the solid waste. The solid waste phase and its attendant moisture storage capacity will determine the quantity and time of the first appearance of the leachate. Therefore, a water balance analysis is first performed on the soil cover to determine the amount of percolation. Then, the solid waste layer is analyzed in relation to the percolation amounts to determine the extent of potential leachate generation.

Treating the moisture flow from the soil cover as a one dimensional system, the water balance model is utilized to calculate the percolation of water into the solid waste. In applying the method, the surface conditions of the sanitary landfill site are defined. The type and thickness of the cover soil, the presence or absence and type of vegetative cover, and the topographical features are the primary surface conditions that affect the percolation. The water balance model is expressed by Equations (5.1)–(5.4). The application of these equations has been demonstrated by Fenn et al. (1975), Dass et al. (1977), and Lu et al. (1984).

$$(PERC) = P - (R/O) - (AET) - (\Delta ST) \tag{5.1}$$

$$I = P - (R/O) \tag{5.2}$$

$$(APWL) = \Sigma NEG(I - PET) \tag{5.3}$$

$$(AET) = (PET) + [(I - PET) - (\Delta ST)] \tag{5.4}$$

The parameters and procedure for conducting the water balance are described as follows:

(1) P = precipitation. This is the mean monthly value based on many years of data obtained from historical records, maintained by the concerned state and federal agencies.

(2) PET = potential evapotranspiration. This is the mean monthly value based on many years of data obtained from the historical records maintained by state and federal agencies. The mean precipitation and

evaporation data, and soil water relationships for specific area are used to develop potential evapotranspiration values. These values may also be derived from Thornthwaite's *PET* equation and associated tabular data developed for a 25 year period (Thornthwaite and Mather, 1957). The procedure for estimating *PET* is as follows: (a) obtain mean monthly air temperature, (b) obtain heat index values for each month corresponding to mean monthly temperature (these values are given in Tables 1 and 2, and Section I of the above reference), (c) obtain from Table 4, Section II of the above reference the unadjusted daily *PET* values at corresponding temperatures and yearly heat index value, (d) make adjustment for these monthly values of unadjusted *PET* by multiplying with the proper correction factors for the given area. This will give the adjusted (*PET*) value for each month. The correction factors for each month with respect to the latitude of the area are given in Tables 6 and 7, Section III of the above reference.

(3) (*R/O*) = surface runoff. This is the monthly mean surface runoff value obtained by multiplying the mean monthly precipitation value by the selected runoff coefficient ($C_{R/O}$). This represents the amount of precipitation that runs off the landfill surface before it can infiltrate into the soil cover.

(4) I = infiltration. It represents the amount of precipitation that enters the surface of the cover soil.

(5) (*I-PET*) = infiltration minus potential evapotranspiration. It is used to determine periods of moisture excess and deficiency in the soil, and is necessary to obtain the difference between infiltration and potential evapotranspiration. A negative value of (*I-PET*) indicates the amount by which the infiltration fails to supply the potential water need of a vegetated area. A positive value of (*I-PET*) indicates the amount of excess water which is available during certain periods of the year for soil moisture recharge and percolation. In most locations there is only one so called "wet" season and one "dry" season per year. Thus, there will be only one set of consecutive negative and one set of positive differences. In dry areas, where precipitation is not sufficient to bring the soil moisture back up to its maximum value of water holding capacity at any given time during the year, the (*I-PET*) value is negative. At locations with positive annual values of (*I-PET*), the soil moisture at the end of the wet period is always at the maximum value of water holding capacity.

(6) (*APWL*) = the accumulated potential water loss. This is the negative value of (*I-PET*), representing the potential water losses that are summed month by month. In most humid areas, since the sum of all

. the (*I-PET*) values is positive, the value of accumulated potential water loss (*APWL*) is assigned zero to the month preceding the first negative (*I-PET*) value. The reason for this is that the soil moisture at the end of the wet season is at field capacity. However, for dry areas [defined as areas where the annual total (*I-PET*) is negative], soil moisture at the end of the wet season is below field capacity. Therefore, it is necessary to find an initial value of (*APWL*) with which to start accumulating the negative values of (*I-PET*). This is done by utilizing Thornthwaite's method of successive approximations (Thornthwaite and Mather 1955).

(7) *ST* = soil moisture storage. This factor represents the soil moisture, or the moisture retained in the soil after a given amount of accumulated potential water loss or gain has occurred. The initial value is calculated at field capacity by multiplying available water per unit depth of soil by root zone depth. This initial value of *ST* is assigned to the last month having a positive value of (*I-PET*), i.e., the last month of the wet season. In dry areas, soil moisture at the end of the wet season is below field capacity. Thus, the initial as well as subsequent *ST* values must be determined from the appropriate soil moisture retention table, utilizing the values of (APWL) calculated as above.

To determine the soil moisture retained each month, Thornthwaite and Mather (1955) have developed soil moisture retention tables for various water holding capacities of various soils. After the soil moisture storage for each of the months with negative values of (*I-PET*) has been found, the positive values of (*I-PET*), representing additions of moisture to the soil, must be added to the previous month's *ST* value. No *ST* value can exceed soil moisture storage at field capacity. Thus, any excess of (*I-PET*) above this maximum *ST* value becomes percolation.

(8) (Δ*ST*) = change in soil moisture storage. It represents the change in soil moisture from month to month.

(9) (*AET*) = actual evapotranspiration. This represents the actual amount of water loss during a given month. As soil moisture is depleted, the rate of evapotranspiration decreases below its potential rate, thereby resulting in an (*AET*) value less than the corresponding *PET* value. For those months where (*I-PET*) is positive, the rate of evapotranspiration is not limited by moisture availability, and (*AET*) is equal to *PET*. For those months where (*I-PET*) is negative, the rate of evapotranspiration is limited by soil moisture availability, and expressed by Equation (5.4).

(10) (*PERC*) = percolation. After the soil moisture storage reaches its maximum, any excess infiltration becomes percolation through the

cover soil and into the underlying solid waste. Therefore, significant percolation will occur only during those months when I exceeds *PET*, i.e., (*I-PET*) is positive, and the soil moisture exceeds its maximum value. For most humid areas, this will occur during the wet season. For dry areas, significant percolation may never occur.

5.3.3.1 Illustrative Example

To best illustrate the water balance calculations of a sanitary landfill, an example is used here. This example was presented earlier by Fenn et al. (1975), for a humid climate with a sandy type soil; and Orlando, Florida data was used. Fenn et al. (1975) also provided examples in which Cincinnati, Ohio, and Los Angeles, California data were used. Lu et al. (1984) provided 10 solved examples of runoff and moisture routing calculations for different conditions of climatic and soil conditions. Dass et al. (1977) showed moisture balance calculations for a sanitary landfill in Wisconsin. The water balance analysis in the illustrative example is simplified by the following basic assumptions (Fenn et al. 1975).

(1) The type of soil is sandy loam having a field capacity of 200 mm/m, wilting point is 50 mm/m and hence the available water is 150 mm/m.

(2) The landfill has been completed with 0.6 m thick final cover and graded with a 2 to 4 percent slope over most of the surface area. Therefore, the soil moisture storage is 150 mm/m × 0.6 m = 90 mm at field capacity. Since there is no soil moisture retention table for 90 mm, the soil moisture retention table for 100 mm is used (Table A-1, Appendix A).

(3) The assumed soil surface runoff coefficient is 0.075 for all months. This value is based on grass and sandy soil at 2% slope (Table 5.2).

(4) The solid waste, cover soil, and vegetative cover were emplaced instantaneously at the beginning of the first month of the computation initiation. Thus, any percolation that may occur prior to the placement of the final cover soil is ignored.

(5) The final use of the site is an open green area to be used for recreation or pasture.

(6) The surface is fully vegetated with a moderately deep-rooted grass, the roots of which draw water directly from all parts of the soil cover but not from underlying solid waste.

(7) The sole source of infiltration is precipitation falling directly on the surface of the landfill. All surface runoff from adjacent drainage areas is diverted around the landfill surface. All ground water infiltration is prevented through proper site selection and design.

(8) The hydraulic characteristics of the soil cover and compacted solid waste are uniform in all directions.

(9) The depth of the landfill is much less than its length and width. Therefore, the water movement is vertically downward.

The water balance calculations for the soil cover are depicted in Table 5.4 and Figure 5.7.

Following is a summary of the water balance for the example problem. The precipitation is sufficiently greater than the evapotranspiration to exceed the soil moisture storage capacity and produce percolation. The fluctuating nature of percolation will cause variations in leachate generation. Following is the summary information obtained from Table 5.4.

Mean annual precipitation, P = 1342 mm
Mean annual runoff, (R/O) = 100 mm
Mean annual infiltration, I = 1242 mm
Mean annual actual evapotranspiration, (AET) = 1172 mm
Mean annual potential evapotranspiration, PET = 1206 mm
Mean annual percolation, $(PERC)$ = 70 mm

5.3.4 SIMPLIFIED METHOD

Moisture routing through a sanitary landfill can be conducted using a simplified method based on precipitation, runoff, and potential evapotranspiration. The moisture deficit is calculated for the period, and the cumulative value gives the water percolation. In this simplified method, the moisture routing calculation is done for two consecutive years. The difference between the second and the first year's accumulated excess water gives the water percolation per year. During subsequent years the same value of percolation will occur. An example is shown below to illustrate the simplified procedure.

5.3.4.1 Illustrative Example

The data from the previous example are used to show the calculation steps for the simplified method. It is assumed that the field capacity of the soil is 200 mm/m, initial soil moisture content of the soil cover is 180 mm/m, and wilting point is 50 mm/m. Therefore, the

Maximum cover moisture deficit = (200 − 50) mm/m × 0.6 m = 90 mm

Initial cover moisture deficit = (200 − 180) mm/m × 0.6 m = 12 mm

The calculation steps are summarized in Table 5.5. The calculations in

TABLE 5.4. Water Balance Data for the Example Problem.

Parameter[a]	J	F	M	A	M	J	J	A	S	O	N	D	Annual
PET	33	39	59	90	140	167	175	173	142	100	53	35	1206
P	50	56	91	88	81	161	230	180	200	121	39	45	1342
($C_{R/O}$)	.075	.075	.075	.075	.075	.075	.075	.075	.075	.075	.075	.075	
(R/O)	4	4	7	7	6	12	17	13	15	9	3	3	100
I	46[b]	52	84	81	75	149	213	167	185	112	36	42	1242
(I − PET)	13[c]	13	25	−9	−65	−18	38	−6	43	12	−17	7	36
(APWL)			(0)[d]	−9[e]	−74[f]	−92	−25[g]	−31[h]	0	0	−17	0	
ST, TABLE A-1 APPENDIX A	100	100	100	91	47	39	77	73	100	100	84	91	
(ΔST)	+9[i]	0	0	−9[j]	−44	−8	38	−4	27[k]	0	−16	+7	
(AET)	33[l]	39	59	90[m]	119	157	175	171	142	100	52	35	1172
(PERC)	+4[n]	13	25	0	0	0	0	0	16	12	0	0	70

[a] The parameters are as follows: PET, potential evapotranspiration; P, precipitation; ($C_{R/O}$), surface runoff coefficient; (R/O), surface runoff; I, infiltration; ST, soil moisture storage; (ΔST), change in storage; (AET), actual evapotranspiration; (PERC), percolation; (APWL), accumulated potential water loss. All values are in millimeters (1 inch = 25.4 mm).
[b] 46 = (50 − 4)
[c] 13 = (46 − 33)
[d] The soil moisture is at the maximum.
[e] −9 = (0 + (−9))
[f] −74 = −9 + (−65)
[g] The situation where a positive (I − PET) value occurs between two negative values is a special case. Here, ST is found by direct addition of (I − PET) to the preceding ST. The (APWL) value is then found from the soil moisture retention table (Table A-1, Appendix A) for the ST value, (+38 + 39 = 77). This value is −25.
[h] −31 = −25 + (−6)
[i] +9 = 100 − 91
[j] −9 = 91 − 100
[k] 27 = 100 − 73
[l] Since (I − PET) is positive (+13), (AET) = PET (33).
[m] (AET) = PET + [(I − PET) − (ΔST)]. 90 = 90 + [(−9) − ((−9)]. 90 = 90 + [−9 − (−9)] or AET = I + (ΔST) = (81 + 9 = 90). Both these procedures for (AET) apply when I is less than PET. When I is greater than PET, (AET) = PET.
[n] 4 = 50 − 4 − 33 − 9
Source: After Fenn et al. (1975).

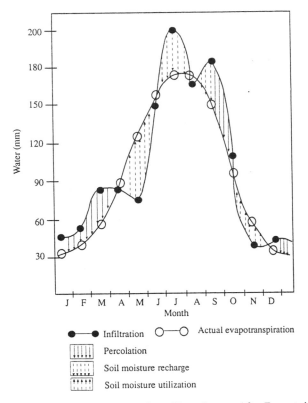

Figure 5.7 Water balance for the example problem. Source: After Fenn et al. (1975).

Table 5.5 show that at the end of 2 years, the excess water reaches 75 mm. During the second year, the annual percolation of 38 mm (77 mm − 39 mm = 38 mm) of excess water will infiltrate into the refuse of the landfill.

5.3.5 LEACHATE GENERATION

Knowing the amount of water that percolates through the cover material, an analysis of the water routing through the solid waste can now be performed to determine the magnitude and timing of leachate generation. An example calculation for leachate generation will be shown here using the moisture routing calculations given in Table 5.4.

Like its cover material, the underlying solid waste cells (including the relatively thin layers of daily cover material) will exhibit a certain capacity to hold water. Qasim (1965), Remson et al. (1968), Qasim and Burchinal (1970a, 1970b), and Korfiatis and Demetracopoulos (1984) determined the field capacity of the solid waste that varied from 20 percent to as high

TABLE 5.5. Moisture Routing Using Simplified Method.

Month	Precipitation[a] (mm)	Runoff (mm)	Infiltration (mm)	Potential Evapotranspiration[d] PET(mm)	Net Water mm/month	Cover Moisture Deficit (mm)	Accumulated Excess Water (mm)
Jan	50	4[b]	46[c]	33	13[e]	$[12-13 = -1]^f$	1
Feb	56	4	52	39	13	−13	14
March	91	7	84	59	25	−25	39[g]
April	88	7	81	90	−9	9	39
May	81	6	75	140	−65	74	39
June	161	12	149	167	−18	92(90)[h]	39
July	230	17	213	175	38	52	39
Aug	180	13	167	173	−6	58	39
Sept	200	15	185	142	43	15	39
Oct	121	9	112	100	12	3	39
Nov	39	3	36	53	−17	20	39
Dec	45	3	42	35	7	13	39
Jan	50	4	46	33	13	0	39
Feb	56	4	52	39	13	−13	52
March	91	7	84	59	25	−25	77[i]
April	88	7	81	90	−9	9	77

TABLE 5.5. (continued).

Month	Precipitation[a] (mm)	Runoff (mm)	Infiltration (mm)	Potential Evapotranspiration[d] PET(mm)	Net Water mm/month	Cover Moisture Deficit (mm)	Accumulated Excess Water (mm)
May	81	6	75	140	−65	74	77
June	161	12	149	167	−18	92(90)[h]	77
July	230	17	213	175	38	52	77
Aug	180	13	167	173	−6	58	77
Sept	200	15	185	142	43	15	77
Oct	121	9	112	100	12	3	77
Nov	39	3	36	53	−17	20	77
Dec	45	3	42	35	7	13	77

[a]Precipitation taken from Table 5–4.

[b]Runoff = Precipitation × assumed runoff coefficient for all months = 50 mm × 0.074 = 4 mm.

[c]Infiltration = Precipitation-runoff = 50 mm − 4 mm = 46 mm.

[d]Potential Evapotranspiration. These values may be developed from the data available from the concerned state or federal agencies for the specific area. Also, the Thornthwaite's (PET) equation and associated tabular data (Thornthwaite and Mather 1955), may be used. In this example values given in Table 5.4 are utilized.

[e]Net water = Infiltration − (PET) = 46 mm − 33 mm = 13 mm.

[f]13 mm of moisture content is percolating into the soil cover, since cover moisture deficit is 12, a minus one (−1) value for deficit is assigned.

[g]No change in accumulated excess water of 39 mm due to cover moisture deficit through the remaining months of the first year.

[h]Maximum cover moisture deficit value is used.

[i]No change in accumulated excess water of 77 mm due to cover moisture deficit through the remaining months of second year.

121

as 35 percent by volume. In other words, the field capacity would vary from about 200 mm water per meter refuse (2.4 inches per foot) to about 350 mm water per meter (4.2 inches per foot). For present illustration, a value of 300 mm per meter (3.6 inches per foot) will be used.

The amount of water which can be added to the solid waste before reaching field capacity depends also on its moisture content when delivered to the landfill site. This value will vary over a wide range depending on the composition of the waste and the climate. Merz and Stone (1970), Fungaroli (1971), Korfiatis and Demetracopoulos (1984), Pfeffer (1992), Tchobanoglous et al. (1993), and others performed moisture analysis on municipal solid waste and showed its moisture content to range anywhere from 10 to 20 percent by volume. A moisture content of 15 percent by volume or about 150 mm per m will be used here. Therefore, with a field capacity of 300 mm per m and an initial moisture content of 150 mm per m the compacted waste would have an absorption capacity of about 150 mm of water per meter of solid waste.

Theoretically, the water movement through a compacted solid waste will act like water movement through a soil layer. In other words, the field capacity of a given solid waste must be exceeded before any significant leaching to a lower level will occur. For the example problem (Table 5.4),

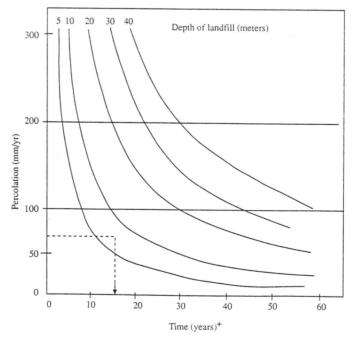

Figure 5.8 Time of first appearance of leachate. Source: After Fenn et al. (1975).

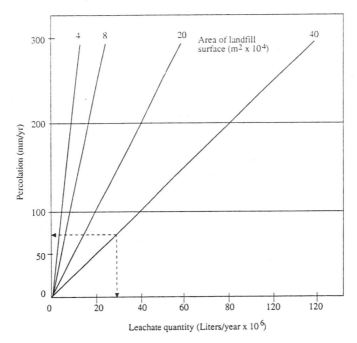

Figure 5.9 Annual leachate quantities after time of first appearance. Source: After Fenn et al. (1975).

this means that 150 mm of percolation would have to be applied to a municipal solid waste layer one meter deep before any significant leachate would be generated from the bottom of that layer. Practically speaking, due to the heterogeneous nature of the solid waste, some channeling of water will develop causing some leaching to occur prior to attainment of field capacity. However, this amount should be small, certainly not a continuous flow, and is assumed negligible.

Employing the above concepts, the extent of the leachate movement for a given sanitary landfill site can be assessed. The time of first appearance of leachate will be influenced by the depth of the landfill and the net percolation. Figure 5.8 shows the relationship between annual percolation amounts and time of first appearance of leachate for various landfill depths. For a mean annual percolation of 70 mm, a 7-m deep landfill having a moisture absorption capacity of 150 mm per m will take 15 years for leachate generation. Figure 5.9 shows the relationship between annual percolation amounts and leachate quantities for various size landfills. A landfill having a surface area of 40 hectare (40×10^4 m²) will produce a leachate quantity of 28×10^6 liters per year if net percolation is 70 mm per year.

5.4 FIELD APPLICATION

The water balance method will serve as a useful engineering tool in conducting environmental assessments of proposed or existing sanitary landfill sites, specifically with regard to leachate generation. However, it should be noted that the method presented in this chapter is intended only to provide a basic tool for estimating the leachate quantity. Certain site specific assumptions will be necessary to utilize the method for a particular location. These assumptions will involve collection of precipitation data, and choice of proper methods for predicting evapotranspiration and runoff from the landfill surface during the operating life of the landfill, and during post-closure care.

The water balance method points out the following characteristics of leachate generation from a sanitary landfill:

(1) Leachate will be generated in humid areas, while no significant amounts may be generated in dry areas.

(2) Leachate generation is not likely to result as a constant flow throughout the year or from year to year, but will follow a pattern somewhat similar to that of precipitation.

(3) In humid areas where leachate will be generated, the hydrogeology of the site will be carefully evaluated to determine its inherent potential for groundwater pollution. Also, for leachate containment, collection and treatment facilities, the quantity can be calculated using the above procedure.

(4) Leachate generation can be minimized by proper and efficient covering operations (including impervious layer), careful contouring and drainage design of the final cover, proper selection of a vegetative cover, and, in some cases, the ultimate use, of the site.

(5) Leachate generation will eventually cease if the final use of the landfill prevents percolation.

It is therefore obvious that leachate will be generated for a long period of time unless percolation is prevented by the closure procedures and final land use. If the water percolation is prevented, the leachate generation will cease shortly after the landfill is completed.

Due to tedious procedures involved in the calculation of leachate generation factors and moisture routing data, it will be beneficial to utilize a suitable computer simulation model. Several simulation models have been discussed in Sec. 5.3.2. With the aid of computer simulation, numerous conditions of soil cover, land use and hydrological factors may be evaluated to integrate the uses of the completed landfill, and minimize percolation and leachate generation. Also, the capacity of leachate treatment plants may be established under various hydrological conditions.

5.5 REFERENCES

Bagchi, A. 1990. *Design, Construction, and Monitoring of Sanitary Landfill*, New York: John Wiley & Sons.

Brunner, D. R. and D. J. Keller. 1972. "Sanitary Landfill Design and Operation," U.S. Environmental Protection Agency, Report SW-65ts, Washington, D.C.: U.S. Government Printing Office.

Chow, V. T., D. R. Mardment and L. W. Mays. 1988. *Applied Hydrology*, New York: McGraw-Hill Book Co.

Dass, P., G. R. Tamke and C. M. Stoffel. 1977. "Leachate Production at Sanitary Landfill Sites," *Journal of the Environmental Engineering Division, Proceedings of the American Society of Civil Engineers*, 103(EE6):981–989.

Fenn, D. G., K. J. Hanley and T. V. DeGeare. 1975. "Use of the Water Balance Method for Predicting Leachate Generation From Solid Waste Disposal Sites," U.S. Environmental Protection Agency, SW-168.

Flanagan, K. 1988. "Methane Recovery Does More Than Provide Energy," *Solid Waste and Power*, 2(4):30–33.

Fungaroli, A. A. 1971. *Pollution of Subsurface Water by Sanitary Landfills*, Vol. 1, Washington, D.C.: U.S. Government Printing Office, p. 200.

Garner, R. B. and E. T. Conrad. 1986. "The Use of the HELP Model in Evaluating Alternative Leachate Management Plans for Three New York City Landfills," *Proceedings of the Waste Tech Conference*, October 20–22, 1986, Washington, D.C.: National Solid Wastes Management Association, and Waste Age Magazine, pp. 1–21.

Gaudy, A. F. and E. T. Gaudy. 1988. *Elements of Bioenvironmental Engineering*, San Jose, CA: Engineering Press, Inc.

Jens, S. W., H. L. Cook, R. A. Hertzler, R. E. Horton and G. W. Musgrave. 1949. "Infiltration," in *Hydrology Handbook, Manual of Engineering Practice No. 28*, New York: American Society of Civil Engineers, p. 33–63.

Joint Committee of the Water Pollution Control Federation and the American Society of Civil Engineers. 1970. "Design and Construction of Sanitary and Storm Sewers," *WPCF Manual of Practice, No. 9*, (ASCE Manual of Practice 37), Washington, D.C.: Water Pollution Control Federation.

Korfiatis, G. P., A. C. Demetracopoulos, E. L. Bourodimos and E. G. Nawy. 1984. "Moisture Transport in a Solid Waste Column," *Journal of Environmental Engineering Division, American Society of Civil Engineers*, 110(4):780–796.

Leckie, J. O., J. G. Pacey and C. Halvadakis. 1979. "Landfill Management with Moisture Control," *Journal of the Environmental Engineering Division, American Society of Engineers*, 105(EE2):337–355.

Linsley, R. K. and J. B. Franzini. 1979. *Water Resources Engineering*, Third Edition, New York: McGraw-Hill Book Co.

Lu, J. C. S., R. D. Morrison and R. J. Stearns. 1981. "Leachate Production and Management from Municipal Landfills: Summary and Assessment," in *Land Disposal of Municipal Solid Waste, 7th Annual Res. Symp.*, EPA-600/9-81/002a, D. W. Shultz, ed., Cincinnati, OH, U.S. Environmental Protection Agency, pp. 1–17.

Lu, J. C. S., R. D. Morrison and R. J. Stearns. 1984. "Production and Management of Leachate from Municipal Landfills: Summary and Assessment," U.S. Environmental Protection Agency EPA-600/2-84-092, Cincinnati, OH: Municipal Environment Laboratory.

Lu, J. C. S., B. Eichenberger and R. J. Stearns. 1985. "Leachate from Municipal Landfills, Production and Managenent," Park Ridge, NJ: Noyes Publ.

Lutton, R. J., G. K. Regan and L. W. Jones. 1979. "Design Construction of Covers for Solid Waste Landfills," U.S. Environmental Protection Agency, EPA-600/2-79/165, Cincinnati, OH: Municipal Environmental Laboratory.

Merz, R. C. and R. Stone. 1970. "Special Studies of a Sanitary Landfill," Springfield, VA: U.S. Department of Health, Education, and Welfare, PB-196-148, p. 222.

Peavy, H. S., D. R. Rowe and G. Tchobanoglous. 1985. *Environmental Engineering,* New York: McGraw-Hill Publishing Co.

Peyton, L. and P. R. Shroeder. 1988. "Field Verification of HELP Model for Landfills," *Journal of Environmental Engineering Division, American Society of Civil Engineers,* 114(EE2):247–269.

Perrier, E. R. and A. C. Gibson. 1980. "Hydrologic Simulation on Solid Waste Disposal Sites," EPA/SW-868, Cincinnati, OH: U.S. Environmental Protection Agency, p. 111.

Pfeffer. J. T. 1992. *Solid Waste Management Engineering,* Englewood Cliffs, NJ: Prentice Hall.

Qasim, S. R. 1965. "Chemical Characteristics of Seepage Water from Simulated Landfills," Ph.D. Dissertation, West Virginia University, Morgantown, WV: p. 145.

Qasim, S. R. and J. C. Burchinal. 1970a. "Leaching from Simulated Landfills," *J. Water Pollution Control Federation,* Vol. 43(3):371–379.

Qasim, S. R. and J. C. Burchinal. 1970b. "Leaching of Pollutants from Refuse Beds," *Journal Sanitary Engineering Division, Proceedings of the American Society of Civil Engineers,* 96(SA-1):49–58.

Remson, I., A. A. Fungaroli and A. W. Lawrence. 1968. "Water Movement in an Unsaturated Sanitary Landfill," *Journal of the Sanitary Engineering Division, Proceedings of the American Society of Civil Engineers,* 94(SA2):307–317.

Schroeder, P. R., J. M. Morgan, T. M. Walski and A. C. Gibson. 1984. "The Hydrologic Evaluation of Landfill Performance (HELP), Model," Technical Resource Document, U.S. EPA PB-85-100840 and PB-85-100832, Cincinnati, OH.

Stanforth, R., R. Ham and M. Anderson. 1979. "Development of a Synthetic Municipal Landfill Leachate," *Journal Water Pollution Control Federation,* 51(7):1965–1975.

Tchobanoglous, G., H. Theisen and S. Vigil. 1993. *Integrated Solid Waste Management: Engineering Principles Management Issues,* New York: McGraw-Hill, Inc.

Thornthwaite, C. W. and J. R. Mather. 1955. "Instructions and Tables for Computing Potential Evapotranspiration and Water Balance," *Publications in Climatology,* Centerton, NJ: Drexel Institute, 10(3):86.

Thornthwaite, C. W. and J. R. Mather. 1957. "Instructions and Tables for Computing Potential Evapotranspiration and Water Balance," *Publications in Climatology,* Centerton, NJ: Drexel Institute, 10(3):185–311.

U.S. Congress Office of Technology Assessment. 1989. *Facing America's Trash: What's Next for Municipal Solid Wastes,* OTA-0-424, Washington, D. C.: U.S. Government Printing Office.

U.S. Environmental Protection Agency. 1992. "MULTIMED Model Documentation," Version 1.01, Center for Exposure Assessment Modeling, Athens, Georgia.

U.S. Soil Conservation Service. 1972. "Hydrology," *SCS National Engineering Handbook,* Sec. 4, Washington, D.C.: U.S. Department of Agriculture.

U.S. Soil Conservation Service. 1975. "Urban Hydrology for Small Watersheds," Technical Release No. 55, Washington, D. C.: U.S. Department of Agriculture, Engineering Division.

Vasuki, N. C. 1988. "Why Not Recycle the Landfill," *Waste Age,* 19(11):165–170.

Viessman, W., J. W. Knapp, G. L. Lewis and T. E. Harbough. 1977. Introduction to Hydrology, Second Edition, New York: IEP-Dun-Donnelly, Harper, and Row Publishers.

Veihmeyer, F. G. 1964. "Evapotranspiration," Section II in *Handbook of Applied Hydrology,* V. T. Chow, ed., New York: McGraw-Hill Book Co., pp. (11-1)-(11-38).

Woodwell, G. M. 1984. *The Role of Terrestrial Vegetation in the Global Carbon Cycle,* New York: John Wiley & Sons.

CHAPTER 6

Leachate Characterization

6.1 INTRODUCTION

PRIOR to 1965 very few people were aware of the fact that water passing through solid waste in a sanitary landfill would become highly contaminated. This water, termed *leachate*, was generally not a matter of concern because few cases of water pollution were noted where leachate had caused harm (Boyle and Ham 1974). It is now known that leachates from sanitary landfills may be an important source of ground water pollution. Walker (1969), and Kelly (1976) reported that many contaminants released from a sanitary landfill, if allowed to migrate, may pose a severe threat to surface and groundwater. Federal, as well as state and local governmental regulations now call for effective control of leachate from sanitary landfills. Effective pollution control requires an understanding of the quantity and quality of landfill leachate.

In this chapter, the literature concerning both experimental investigations, and modeling of leachate quality and leachate characterization are reviewed. The goal is to identify the types and concentration histories of pollutants present in leachate, and to observe relationships among climate, age, refuse placement, and overall landfill stabilization.

6.2 FACTORS AFFECTING LEACHATE QUALITY

Lu et al. (1981, 1984, 1985), Chian and Dewalle (1976), and Chian (1977) provided a detailed literature reviews of landfill composition. These researches indicated that leachate quality was highly variable. Leachate contains larger pollutant loads than raw sewage or many industrial wastes. The variation in leachate quality are generally attributed to a myriad of interacting factors such as type and depth of solid waste, age of fill, the rate of water application, landfill design and operations, and the interaction of leachate with its environment. The effects of some of these variables upon leachate quality are presented below. The quality variations can also be attributed to sampling procedures, sample preservation, handling and storage, and analytical methods used to characterize the leachates.

6.2.1 PROCESSED REFUSE

Leachate characteristics from shredded or baled refuse fills also differ greatly. Fungaroli and Steiner (1979), Kemper and Smith (1981), and Savage and Trezek (1980) conducted experiments with processed landfills (shredded and baled). The results clearly indicated that leachate from shredded fills has significantly higher concentrations of pollutants than those from unshredded refuse. Attainment of field capacity is also delayed, but the rate of pollutant removal, solid waste decomposition rate, and the cumulative mass of pollutants released per unit volume of leachate is significantly increased when compared with unshredded fills.

Baling, however, has shown opposite results on leachate generation and quality. Baling results in a large volume of dilute leachate with a longer period of stabilization than required for unbaled refuse. Once the field capacity of shredded and baled refuse is reached, in the long-run, the cumulative mass of pollutant removal per kg of solid waste will be the same regardless of waste processing (Lu et al. 1984).

6.2.2 DEPTH OF REFUSE

One of the earlier studies performed by Qasim and Burchinal (1970a, 1970b) reported that substantially greater concentrations of constituents are obtained in leachates from deeper fills under similar conditions of precipitation and percolation. Deeper fills, however, require more water to reach saturation, require longer time for decomposition, and distribute the bulk of extracted material over a longer period of time. Water entering from the surface of the landfill and travelling down through the refuse will successively come in contact with solid waste, and the polluting chemicals will successively transfer to the percolating water. Deep fills offer greater contact time and longer travel distance, thus higher concentrations will result. Phelps (undated) also demonstrated higher concentrations of contaminants in the leachate emanating from deeper columns.

6.2.3 CODISPOSAL WITH SEWAGE SLUDGE

Codisposal of municipal solid waste and sludge from municipal wastewater treatment plants has a significant effect upon the generation and quality of leachate. Codisposal has been extensively researched for over three decades. Stone (1974), Emcon Associates (1974), Pohland (1975), and Lu et al. (1984) investigated the effects of codisposal upon the leachate quality. Both septage and sludge may increase the moisture input thus enhancing leachate generation. Microbiological seeding and nutrients enrichment of municipal solid wastes with septage and sludge has increased the rate of biological stabilization (Pohland 1975). Accelerated methanogenic activity was also observed (Emcon Associates 1974). With the exception of more acidic leachate and higher BOD_5, the chemical quality of leachate did not change significantly with codisposal. Lu et al. (1982) reported that codisposal posed no more a pollution threat than the leachate from municipal landfill. Nitrate and enteric pathogens appear to represent the greatest increase in leachate from codisposal activities. Levine and Rear (1989) evaluated the difference in the characteristics of codisposal, hazardous and landfill leachate. These researchers found that the concentrations of organics, dissolved solid metals and inorganics are considerably higher in hazardous landfill leachate than in codisposal or sanitary landfill leachate.

Sludge that is codisposed with municipal solid waste in a sanitary landfill will not have to meet numerical pollutant limits as with other sludge disposal practices regulated under EPA guidelines (U.S. EPA 1993). It is difficult at this time to predict the chemical interactions and the specific effects of the sludge on the landfill environment. EPA, however, believes that many other environmental protection measures specified for landfill design under Subtitle D regulations are sufficient to protect the water environment and human health (Black and Veatch, Inc. 1992).

6.2.4 CODISPOSAL WITH HAZARDOUS WASTES

Municipal landfills may receive hazardous wastes if not managed properly. Pohland et al. (1990) conducted pilot-scale simulated landfill investigations to identify the various attenuating mechanisms operative during landfill codisposal of solid waste and hazardous wastes. These investigators reported that these mechanisms are dependent on the physical-chemical properties of the wastes, their mobility in the prevailing transport media, and the overall characteristics of the landfill environment established during the sequential phases of waste stabilization. The results of these investigations have indicated that landfills possess a finite capacity to attenuate both organic and inorganic hazardous waste constituents when they are codisposed with municipal refuse. The inorganic heavy metals could be attenuated by microbially-mediated physicochemical processes, including reduction, precipitation, sorption and waste matrix containment or encapsulation. Similarly, codisposed organic hazardous waste constituents tended to fractionate according to their physical and chemical properties, and could be attenuated primarily by sorption and biotic or abiotic conversion during the progress of stabilization, with the generation and release of identifiable reaction products. Therefore, although the organic hazardous wastes loaded to the test columns had minor final influence on the progress of landfill stabilization, some of them were more resistant to attenuation. In contrast, heavy metal loadings, capable of inhibiting or delaying methane fermentation and the overall progress of stabilization, could be eventually attenuated and significantly removed from the leachate by capture within the waste matrix. With time, the refractory and less reactive organic and inorganic hazardous wastes may be mobilized and released from the landfill either in leachate or gases. However, the potential release can be minimized by operation control involving gas and leachate containment, collection, utilization or recycle.

6.2.5 CODISPOSAL WITH SORBTIVE WASTES

Studies have shown that codisposal of many sorbtive material such as incinerator ash, fly ash, kiln dust, limestone, and other material with municipal solid waste has some effect upon leachate quality (Liskowitz et al. 1976; Fuller 1978; Chen and Eichenberger 1981). The results indicated that there is a reduction in mobility of many hazardous constituents of leachate. This may be due to (1) adsorption and sorption of metallic ions, (2) formation of less soluble calcium and carbonate compound, and (3) raising of pH may cause precipitation of metals.

6.2.6 AGE OF FILL

Variation of leachate quality with age of fill is expected, because organic matter will continue to undergo stabilization. It should be noted that release of constituents from solid waste is obviously governed by decomposition processes, and the rate and volume of water infiltrating through the fill. Age is merely a convenient means of measuring and monitoring changes in leachate composition, and extraction of pollutants from the refuse bed. As a result, many studies describe leachate quality as a function of time. Many studies have converted time with leachate quantity based on water input rate and leachate generation.

Lu et al. (1984) reviewed the leachate composition data of over 30 investigations conducted over a period of 25 years. The general trend reported is that pollutant concentrations in leachate peak in early life (that is within 2–3 years), followed by a gradual decline in ensuing years. This trend applies to most of the constituents, but in particular to organic indicators (e.g., BOD, COD, TOC, etc.), and microbiological population. Most other constituents exhibit steady decreases in concentration over 3 to

5 years due to continued flushing of the refuse bed. Among these are iron, zinc, phosphate, chloride, sodium, copper, organic nitrogen, total solids, and suspended solids. In some cases however, the concentration of heavy metals fluctuated because of precipitation, dissolution, adsorption and complexation mechanisms that may retain or mobilize the metals within the landfill microenvironments.

Several classes of organic compounds have been identified in landfill leachate. Lu et al. (1984) conducted a comprehensive review of literature on organic components in leachate. These researchers concluded that, in general, the compounds can be classified into three groups: (1) fatty acids of low molecular weight, (2) humic, carbohydrate-like substances of intermediate molecular weight, and (3) fulvic-like substances of intermediate molecular weight. The relative proportions of these components are largely a factor of landfill age. For relatively unstabilized landfills, results show that up to 90 percent of the soluble organic carbon is due to short-chain volatile fatty acids. Among these acids, acetic, propionic, and butyric acids are present in greatest concentrations. The next largest fraction is usually fulvic acids with a relatively high density of carboxyl and aromatic hydroxyl groups. As the landfill ages, the proportions of organic components change. A decrease in volatile fatty acids, and increase in fulvic-like fractions with age of fill has been reported (Lu et al. 1984).

Early research clearly indicate that pathogenic bacteria are present in fresh leachate. Viruses are occasionally detected. There is however, a significant deactivation in bacterial and viral populations with solid waste age. This is attributed to adverse environmental conditions such as initial temperature rise, persistant low pH, and the presence of chemicals that are generally associated with solids decomposition processes.

6.3 EXPERIMENTAL WORK

Qasim and Burchinal (1970a, 1970b), Fungaroli (1971), Rovers and Farquhar (1973), Walsh and Kinman (1979), and Wigh and Brunner (1979) utilized laboratory-scale and pilot testing using columns and lysimeters 0.3–1.8 m diameter and 0.6–3.1 m deep filled with compacted municipal solid waste. Kelly (1976), Hentrich et al. (1979), Leckie et al. (1979), and Wigh (1979) investigated the behavior of pilot or field-scale landfills ranging from 0.9–4.7 m deep. Several other researchers investigated the behavior of experimental landfills operated with leachate recirculation (Pohland, 1975; and Pohland et al. 1979; Birbeck 1980). The methodologies utilized in all these studies were similar. An input/output approach where controlled amounts of water were added, and liquid leachate and gas outputs were monitored. Straub and Lynch (1982a, 1982b), and Chian and DeWalle (1976), provided several generalizations concerning landfill leachate characterization and behavior. Some of these generalizations are given below:

(1) In virtually all experiments, the leachate production rate was delayed because of the moisture absorption capacity of the refuse. The moisture balance on the landfill refuse indicated that, after commencement of significant leachate production, the overall moisture content of the landfill remains essentially constant. Field capacities of refuse computed from input and output moisture data generally ranged between 300 and 400 mm/m.

(2) High concentrations of organic and inorganic contaminants are typically associated with leachate. Peak concentrations of COD and total solids above 50,000 mg/L are common. However, a wide range of concentrations of various contaminants have been observed for different landfills at various ages. Chian (1977), investigated the characteristics of leachate from 13 field and laboratory scale landfills. While wide differences in leachate composition were noted, a meaningful qualitative comparison was made on the basis of landfill age, utilizing the ratios of COD to TOC, BOD to COD, volatile solids (VS) to fixed solids (FS), and percentage of free volatile fatty acid carbon to TOC. These investigators suggested that there is a general decrease in both organic and inorganic strength with age. They characterized young landfills as having high strength leachate, while dilution and microbial utilization of organics reduced leachate strength in older landfills. The pattern of high contaminants concentrations near the onset of leaching, followed by a gradual decrease, is typical among other investigations, although specific times for the decrease vary considerably.

(3) Most of the experimental works utilized similar factors such as physical dimensions of the landfills, refuse compaction and density, soil cover, initial moisture content, the rate of moisture application, and temperature control. All these factors varied greatly. Natural moisture inputs, as well as artificially supplied moisture, have varied from constant application to seasonal inputs. The quantity of input moisture varied from 0.1 cm to 1.0 cm per day. The time span of reported results varied from several weeks to several years. Furthermore, since the specific objectives of these studies differ, the measured variables and the form in which the results are reported have also varied.

6.4 LEACHATE MODELING

Various mathematical models have been developed to estimate the concentration of pollutants in the leachate. Fungaroli (1971), Fungaroli and Steiner (1971), Fenn et al. (1975), Schroeder et al. (1984), and others developed methods to estimate the quantity of leachate production from sanitary landfills. Qasim and Burchinal (1970b) operated three experimental landfill columns of varying heights, and applied column adsorption theory to describe the leaching of chloride. The concentration histories of 14 other contaminants were related to the chloride estimates, with reasonable agreement. The procedure relies on empirically derived parameters, but is responsive to depth of refuse and rate of moisture flow through the fill (Qasim 1965; Qasim and Burchinal 1970a, 1970b; Burchinal 1970). A semi-empirical equation was developed by Wigh and Brunner (1979) to describe the concentration history of various contaminants in leachate generated from an experimental landfill. The equation is based on two consecutive first order reactions and expresses the concentrations of contaminants as a function of cumulative leachate volume, maximum concentration, and two rate constants. Parameters of the equation were evaluated to obtain a good visual fit to concentration histories of several contaminants. Straub and Lynch (1982a, 1982b) indicated that the model captures the general decrease from high initial concentration, which is typical of observed leachate behavior, but its empirical nature limits its use to landfills of similar refuse, density, and depth. Lu et al. (1984) developed a relationship between landfill age and various constituents in leachate. Similarly, Raveh et al. (1979) utilized an experimental function to describe the concentration histories of various pollutants from experimental landfill columns. The declining concentrations of empirically fit parameters were evaluated for one set of experiments. Phelps (undated) developed a model of sanitary landfill leaching utilizing mass transfer equations based on flow through a moisture film in the refuse particles. The model is applied for assumptions of constant moisture infiltration rate and constant moisture content above a wetting front. Model predictions are compared with observed results from several experimental landfill columns which were subjected to various moisture application rates and contained refuse at different depths. Although the model requires the empirical estimation of several parameters and makes limiting assumptions about moisture flow, it is based on descriptions of fundamental mass transfer and contaminant transport processes.

Straub and Lynch (1982a, 1982b), proposed mathematical models for the movement of contaminants in the leachate generated in a sanitary landfill. The first mathematical model is based on simple completely mixed reactor concepts, and on unsaturated flow and transport in porous media (Straub and Lynch 1982a). These researchers also provided computer simulations for laboratory-scale experimental landfills reported by others. General agreement between simulated and observed results indicate that leachate behavior is explainable in terms of fundamental transport processes. The roles of moisture retention in the landfill, and dilution by infiltrating water are of primary importance in understanding and predicting leachate quantity and quality.

The second set of models proposed by Straub and Lynch (1982b) utilize dissolution, transport, and decay of organic materials in unsaturated sanitary landfills. The models are based on cascade reactor concepts and on unsaturated moisture and contaminant transport in porous media. The roles of aerobic and anaerobic bacteria are simulated using conventional kinetic formulations. Computer simulations are obtained for landfill experiments reported by others, and good agreement with observed behavior is demonstrated. The results indicate that aerobic activity is negligible over the leaching life of a landfill, and that anaerobic activity is significantly inhibited.

Demetracopoulos et al. (1986) developed a mathematical model for the generation and transport of solute contaminants through a solid waste landfill. The governing equations are solved numerically by the method of lines. A sensitivity analysis of the model showed that care must be exercised in the selection of the time step and the grid size. The model output agrees quantitatively with measurements of concentration. It however, produces a hydrograph-like contaminant concentration history at the landfill bottom.

6.5 LEACHATE QUALITY

Lu et al. (1984) in an effort to develop a relationship between landfill age and various constituents in leachate, analyzed the data of laboratory field test cells at actual landfill sites. The data was plotted for various constituents. These plots for BOD, COD, TOC, total alkalinity, calcium, potassium, sodium, sulfate and chloride are shown in Figure 6.1. Lu et al. (1984) also developed the upper concentration boundaries using a first order rate equation. In most cases, the first order rate equation conveniently expressed the behavior of the upper limits of the experimental data. These equations are given in Table 6.1. In most cases, the rate equations provide adequate behavior in a limited time range. These equations have application for landfills of age greater than three years and less than thirty years. For several constituents, Lu et al. (1981 and 1984) could not develop equations due to insufficient data or the absence of an apparent trend.

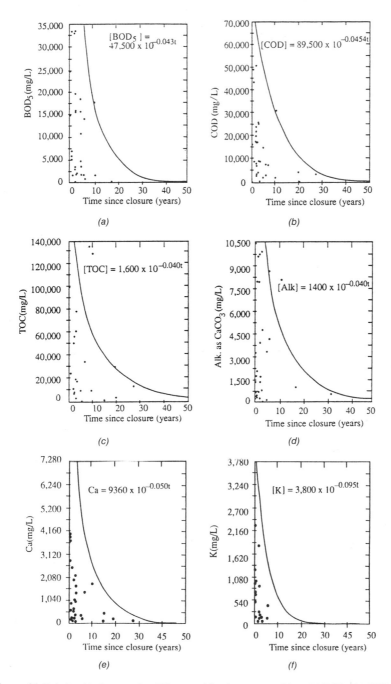

Figure 6.1 Relationship between landfill age and leachate composition: (a) BOD, (b) COD, (c) TOC, (d) alkalinity, (e) calcium and (f) potassium.

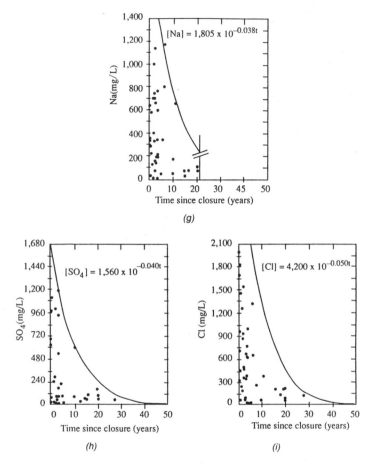

Figure 6.1 (continued) (g) Sodium, (h) sulfate and (i) chloride. Source: Lu et al. (1984).

Many of these constituents are nitrate, total phosphorus, iron, manganese, lead, zinc, and pH.

Chian and DeWalle (1976, 1977a, 1977b) and Chian (1977), summarized the chemical quality of leachate from 30 sanitary landfills. The data was reported by many researchers for landfills of different depths, moisture inflow, and ages. The compositions of leachate from different fills as reported in the literature, show a large variation. The age of landfill and thus the degree of solid waste stabilization has a significant effect on the composition of leachate. Other factors that contribute to the variation in quality are: solid waste characteristics, composition of waste, size of fill and degree of compaction, the moisture content and degree of rainwater infiltration, temperature, sampling, and analytical methods. Other factors such as landfill geometry and interactions of leachate with its environment prior to sample collection, also contribute to the spread of data (DeWalle and Chian 1974; Chian and DeWalle 1977a). The typical composition of three different landfill leachate at ages 1, 5 and 16 years are provided in Table 6.2. Pfeffer (1986), referred to the work of Keenan et al. (1983), who provided the average composition of leachate collected each year over a period of four years (Table 6.3). Tchobanoglous et al. (1993), based on the results of many studies, developed leachate characteristics data for 2 and 10 year old landfills (Table 6.4).

Chian and DeWalle (1976, 1977a), and Chian (1977), also reported that there is a wide reduction with time in the concentrations of organic matter in leachate as measured by COD, TOC and BOD. A COD of 71,670 mg/L was obtained in a leachate from an 8-month-old uncovered landfill containing milled refuse. The COD in the leachate from the same landfill after 17 years was 81 mg/L. The concentrations of inorganics as measured by fixed solids (FS) also fluctuated from a high value of 22,895 mg/L to a low value of 812 mg/L. Likewise, other organic and inorganic contaminants also followed the same trend of decreasing strength with increase in age of

TABLE 6.1. Rate Equations and Rate Constants for Leachate Constituents Concentration Histories.

Rate Equation	Units	Rate Constant, k
$BOD_5 = 47,500 \times 10^{-kt}$	mg/L	0.043[a]
$COD = 89,500 \times 10^{-kt}$	mg/L	0.0454[a]
$TOC = 1,600 \times 10^{-kt}$	mg/L	0.040[a]
$TVS = 24,000 \, e^{-kt}$	mg/L	0.185[b]
$SP\text{-}COND = 20,850 \, e^{-kt}$	mg/L	0.10[b]
$TDS = 16,000 \, e^{-kt}$	mg/L	0.075[b]
$Org\text{-}N = 130 \, e^{-kt}$	mg/L	0.185[b]
$NH_4\text{-}N = 12,000 \, e^{-kt}$	mg/L	0.10[b]
$SO_4 = 15,000 \, e^{-kt}$	mg/L	0.079[b]
$ALK^c = 1,400 \times 10^{-kt}$	mg/L	0.040[a]
$Ca = 9,360 \times 10^{-kt}$	mg/L	0.050[a]
$K = 3,800 \times 10^{-kt}$	mg/L	0.095[a]
$Na = 1,805 \times 10^{-kt}$	mg/L	0.038[a]
$SO_4 = 1,560 \times 10^{-kt}$	mg/L	0.040[a]
$Cl = 4,200 \times 10^{-kt}$	mg/L	0.050[a]
$Cr = 330 \, e^{-kt}$	μg/L	0.90[b]
$Cu = 10 \, e^{-kt}$	mg/L	0.20[b]
$Cd = 160 \, e^{-kt}$	μg/L	0.125[b]

[a]base 10.
[b]base e.
[c]as $CaCO_3$.
Source: Lu et al. (1984).

TABLE 6.2. Composition of Landfill Leachate.

Parameters	Age of Landfill		
	1 Year	5 Year	16 Year
BOD	7,500–28,000	4,000	80
COD	10,000–40,000	8,000	400
pH	5.2–6.4	6.3	
TDS	10,000–14,000	6,794	1,200
TSS	100–700		
Specific Conductance	600–9,000	—	
Alkalinity (CaCO$_3$)	800–4,000	5,810	2,250
Hardness (CaCO$_3$)	3,500–5,000	2,200	540
Total P	25–35	12	8
Ortho P	23–33	—	
NH$_4$-N	56–482		
Nitrate	0.2–0.8	0.5	1.6
Calcium	900–1,700	308	109
Chloride	600–800	1,330	70
Sodium	450–500	810	34
Potassium	295–310	610	39
Sulfate	400–650	2	2
Manganese	75–125	0.06	0.06
Magnesium	160–250	450	90
Iron	210–325	6.3	0.6
Zinc	10–30	0.4	0.1
Copper	—	<0.5	<0.5
Cadmium	—	<0.05	<0.05
Lead	—	0.5	1.0

Note: All values are in mg/L except specific conductance measured as microhms per centimeter and pH as pH units.
Source: After Chian and DeWalle (1976, 1977a).

the fills. These researchers have also shown that the ratio of many chemical constituents such as COD/TOC, BOD/TOC, free volatile fatty acid/TOC, and VS/FS also decrease with the age of the fill. These ratios are shown in Figure 6.2 (Chian and DeWalle 1976, 1977a). It can be seen from Figure 6.2(a) that the COD/TOC ratio tends to decrease as the landfill ages. This ratio varied from 3.3 for a relatively young fill to 1.2 for an old fill. The maximum possible COD/TOC ratio for several organic compounds is 4.0 and it can be as low as 1.3 for organics containing carboxyl groups. A decrease in this ratio in the leachate sample represents a more oxidized state of the organic carbon which becomes less readily available as an energy source for microbial growth. These organics are generally degradation products of microbial activity and increase with the age of landfill. The resulting leachate thus becomes less amenable to biological treatment (DeWalle and Chian, 1974).

The BOD test in general represents the biodegradable portion of total organics present in the leachate. Similar to COD/TOC ratio, the BOD/COD ratio also shows a decrease as the age of the landfills increase. Similar results were obtained in a study conducted by Miller et al. (1974). The calculated ratio of BOD/COD based on Miller's data showed a decrease from 0.47–0.07 within a period of 23 years. When compared to the results of Chian and DeWalle (1976, 1977a), [Figure 6.2(b)], the ratio decreased from 0.8–0.05 within a time span of 17 years.

The ratio of the free volatile fatty acids as a percent of TOC also shows a decrease with the age of the fill [Figure 6.2(c)]. Since the free volatile fatty acids are readily biodegradable, a decrease in the ratio of carbon present in free volatile acids to TOC corresponds with the decrease in BOD/COD ratio. Figure 6.2(c) shows a decrease in the ratio of free volatile fatty acids from 0.49–0.05 with increasing age of the landfill.

The organic compounds in leachate decrease more rapidly than the inor-

TABLE 6.3. Landfill Leachate Characteristics Over Four Year Period.

Item	Year 1	Year 2	Year 3	Overall
BOD_5	4,460	13,000	11,359	10,907
COD	11,210	20,032	21,836	18,533
TSS	1,994	549	1,730	1,044
Dissolved Solids	11,190	14,154	13,181	13,029
pH	7.1	6.6	7.3	6.9
Alkalinity ($CaCO_3$)	5,685	5,620	4,830	5,404
Hardness ($CaCO_3$)	5,116	4,986	3,135	4,652
Calcium	651	894	725	818
Magnesium	652	454	250	453
Phosphate	2.8	2.6	3.0	2.7
Ammonia-N	1,966	724	883	1,001
Kjeldahl-N	1,660	760	611	984
Sulfate	114	683	428	462
Chloride	4,816	4,395	3,101	4,240
Sodium	1,177	1,386	1,457	1,354
Potassium	969	950	968	961
Cadmium	0.04	0.09	0.10	0.09
Chromium	0.16	0.43	0.22	0.28
Copper	0.44	0.39	0.32	0.39
Iron	245	378	176	312
Nickel	0.53	1.98	1.27	1.55
Lead	0.52	0.81	0.45	0.67
Zinc	8.70	31	11	21
Mercury	0.007	0.005	0.011	0.007

Note: All values in mg/L except pH.
Source: Adapted from Keenan et al. (1983) with permission from ASCE.

TABLE 6.4. Typical Data on the Composition of Leachate from New and Mature Landfills.

Constituent	New landfill (less than 2 years)		Mature landfill (greater than 10 years)
	Range[b]	Typical[c]	Range[b]
BOD₅ (5-day biochemical oxygen demand)	2,000–30,000	10,000	100–200
TOC (total organic carbon)	1,500–20,000	6,000	80–160
COD (chemical oxygen demand)	3,000–60,000	18,000	100–500
Total suspended solids	200–2,000	500	100–400
Organic nitrogen	10–800	200	80–120
Ammonia nitrogen	10–800	200	20–40
Nitrate	5–40	25	5–10
Total phosphorus	5–100	30	5–10
Ortho phosphorus	4–80	20	4–8
Alkalinity as $CaCO_3$	1,000–10,000	3,000	200–1,000
pH	4.5–7.5	6	6.6–7.5
Total hardness as $CaCO_3$	300–10,000	3,500	200–500
Calcium	200–3,000	1,000	100–400
Magnesium	50–1,500	250	50–200
Potassium	200–1,000	300	50–400
Sodium	200–2,500	500	100–200
Chloride	200–3,000	500	100–400
Sulfate	50–1,000	300	20–50
Total Iron	50–1,200	60	20–200

[a]Except pH, which has no units.
[b]Representative range of values. Higher maximum values have been reported in the literature for some of the constituents.
[c]Typical values for new landfills will vary with the metabolic state of the landfill.
Source: Adapted from Tchobanoglous, et al. (1993) with permission from McGraw-Hill Book Co.

ganics with increasing age of the landfill. The decrease in organics is due to decomposition and washout. The inorganics only decrease due to washout by infiltrating rainwater. The ratio of total volatile solids to total fixed solids VS/FS also decreases with age. These results are plotted in Figure 6.2(d). There is a gradual decrease from a value of 2.0 for a young fill to 0.2 for old fills. Other researchers found a decrease from 2.1 to 1.9 in a short time (Qasim 1965).

Chian and DeWalle (1976, 1977a), also provided similar plots for sulfate to chloride ratio, oxidation reduction potential (ORP), and pH. These results are plotted in Figure 6.3. These plots also reflect a change in these parameters with the stabilization of landfills. The rapid decrease in SO_4/Cl ratio with time [Figure 6.3(a)] is attributed to a decrease in the concentra-

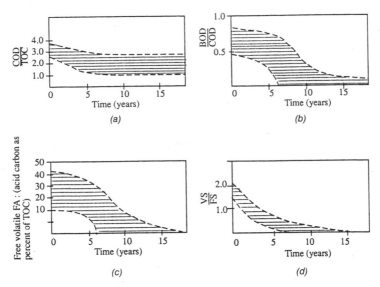

Figure 6.2 Changes in COD/TOC and BOD/COD ratios, percent change in free volatile acid carbon to TOC, and VS/FS ratios with age of fill. Source: Chian and DeWalle (1976 and 1977a).

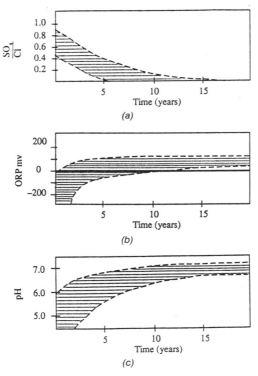

Figure 6.3 Leachate SO_4, Cl, ORP, and pH changes with time in sanitary landfills. Source: Chian and DeWalle (1976 and 1977a).

tion of sulfate as a result of anaerobic conditions prevailing in the landfill in which sulfate is reduced to sulfide. The latter is precipitated out with iron and other heavy metals. The trend in the SO_4/Cl ratio is inversely related to the ORP parameter. The largest decrease in SO_4/Cl ratio corresponds with the lowest ORP values confirming the high degree of anaerobiosis in the landfill. The decrease in ORP readings with time, as shown in Figure 6.3(b) also reflects the degree of stabilization of leachate. Figure 6.3(c) shows the increase in pH of leachate with time, which reflects the decrease in the concentration of the partially ionized free volatile fatty acids. All these studies clearly indicate that (1) leachate quality is highly variable; (2) the quality of leachate will change with age, and therefore, the treatment facility must be flexible or upgraded to treat the changing leachate quality; (3) a reliable estimate of the chemical quality of the leachate can be made using the relationships given earlier; and (4) analytical data should be developed for specific conditions.

6.6 TOXICITY OF MUNICIPAL LANDFILL LEACHATE

In addition to many organic and inorganic compounds (measured by routine water quality tests) that are present in landfill leachates, release of potentially toxic substances have also been reported. The general lack of engineering controls at most existing landfills, combined with the fact that many landfills inadvertently may have accepted in the past, or continue to accept hazardous wastes or industrial nonhazardous wastes has enhanced potentially toxic substances releases. Such releases occur primarily through three pathways: migration of leachate to groundwater, migration of leachate and runoff to surface water, and emission of volatile gases to the atmosphere. Median concentrations of many toxic substances found in municipal landfill leachates are provided in Table 6.5 (U.S. EPA 1988). As a result of high concentrations of many toxic constituents the landfills can be a threat to human health because of their potential for water contamination.

Brown et al. (1991) investigated the acute and genetic toxicity of municipal landfill leachate. In this study the municipal solid waste landfills were found to contain many of the same hazardous constituents as found in hazardous waste landfills. Because of the large number of municipal solid waste landfills, these sites pose a serious environmental threat to groundwater quality. Four leachate and one up-gradient groundwater samples were collected from landfills selected to be representative of landfills of differing ages and type of wastes (Brown et al. 1991). Each sample was tested through three genetic toxicity bioassays (the *Aspergillus* diploid assay, and the *Salmonella*/microsome assay). In addition to the

TABLE 6.5. Median Concentrations of Substances
Found in Municipal Landfill Leachate.

Substance[a]	Median Concentration (ppm)	Substance[a]	Median Concentration (ppm)
Inorganics:		**Organics** *(continued):*	
Antimony (11)	4.52	Dichlorodifluoromethane (6)	237
Arsenic (72)	0.042	1,1-Dichloroethane (34)	1,715
Barium (72)	0.853		
Beryllium (6)	0.006	1,2-Dichloroethane (6)	1,841
Cadmium (46)	0.022	1,2-Dichloropropane (12)	66.7
Chromium (total) (97)	0.175	1,3-Dichloropropane (2)	24
Copper (68)	0.168	Diethyl phthalate (27)	118
		2,4-Dimethyl phenol (2)	19
Cyanide (21)	0.063	Dimethyl phthalate (2)	42.5
Iron (120)	221	Endrin (3)	16.8
Lead (73)	0.162	Ethyl benzene (41)	274
Manganese (103)	9.59	bis (2-Ethylhexyl)	
Mercury (19)	0.002	phthalate (10)	184
Nickel (98)	0.326	Isophorone (19)	1,168
Nitrate (38)	1.88	Lindane (2)	0.020
Selenium (18)	0.012	Methylene chloride (68)	5,352
Silver (19)	0.021	Methyl ethyl ketone (24)	4,151
Thallium (11)	0.175	Naphthalene (23)	32.4
Zinc (114)	8.32	Nitrobenzene (3)	54.7
		4-Nitrophenol (1)	17
		Pentachlorophenol (3)	173
Organics:		Phenol (45)	2,456
Acrolein (1)	270	1,1,2,2-Tetrachloroethane (1)	210
Benzene (35)	221	Tetrachloroethylene (18)	132
Bromomethane (1)	170	Toluene (69)	1,016
Carbon tetrachloride (2)	202	Toxaphene (1)	1
Chlorobenzene (12)	128	1,1,1-Trichloroethane (20)	887
Chloroform (8)	195		
bis (Chloromethyl) ether (1)	250	1,1,2-Trichloroethane (4)	378
p-Crestol (10)	2,394	Trichloroethylene (28)	187
2,4-D (7)	129		
4,4-DDT (16)	0.103	Trichlorofluoromethane (10)	56.1
Di-n-butyl phthalate (5)	70.2	1,2,3-Trichloropropane (1)	230
1,2-Dichlorobenzene (8)	11.8	Vinyl chloride (10)	36.1
1,4-Dichlorobenzene (12)	13.2		

[a]Number of samples in parentheses.
Source: Adapted from U.S. EPA 1988.

three genetic toxicity assays, each sample was tested in the *Microtox* test
to measure acute toxicity. The three bioassays measured the ability of each
sample to induce mutations in bacteria, bind to microbial DNA, or cause
chromosome damage in diploid fungi. Genetically toxic chemicals may
cause cancer, genetic disease, sterility, abortions, heart disease or a vari-

ety of other chronic effects. These chronic effects can be subtle and may not appear for decades after exposure.

The results of the acute and genetic toxicity bioassays combined with the chemical analyses and associated cancer risk assessment clearly showed that leachate from municipal solid waste landfills is just as toxic as leachate from landfills in which residential and hazardous wastes were codisposed (Brown et al. 1991).

6.7 REFERENCES

Birbeck, A. E., et al. 1980. "Evaluation of Landfill Beds as Anaerobic Treatment Systems for Landfill Leachates," *Proceedings of the Technical Association of the Pulp and Paper Industry, 1980 Environmental Conference,* Denver, Colorado.

Black and Veatch. 1992. "Sludge Regulations Update," Issue No. 1. pp. 1–6.

Boyle, W. C. and R. K. Ham. 1974. "Biological Treatability of Landfill Leachate," *Journal Water Pollution Control Federation,* 46(5):860–872.

Brown, K. W., G. E. Schrab and K. C. Donnelly. 1991. "Acute and Genetic Toxicity of Municipal Landfill Leachate, TR-153, Texas Water Resources Institute, Texas A & M University.

Burchinal, J. C. 1970. "Microbiology of Acid Production in Sanitary Landfills," A Summary report, U.S. Department of Health Solid Waste Program Grant No. UI00529.

Chen, K. Y. and B. A. Eichenberger. 1981. "Evaluation of Costs for Disposal of Solid Wastes from High-BTU Gasification," Energy and Environmental Systems Division. Argonne National Laboratory, Argonne, Illinois, pp. 127.

Chian, E. S. K. and F. B. DeWalle. 1976. "Sanitary Landfill Leachates and Their Treatment," *Journal of the Environmental Engineering Division, ASCE,* 103(EE2): 411–431.

Chian, E. S. K. and F. B. DeWalle. 1977a. "Evaluation of Leachate Treatment, Vol. I, Characterization of Leachate, EPA-600/2-77-186a," Cincinnati, Ohio: U.S. Environmental Protection Agency.

Chian, E. S. K. and F. B. DeWalle. 1977b. "Evaluation of Leachate Treatment Vol. II, *Biological and Physical-Chemical Processes,* EPA-600/2-77-186b, Cincinnati, Ohio: U. S. Environmental Protection Agency.

Chian, E. S. K. 1977. "Stability of Organic Matter in Landfill Leachates," *Water Research,* Pergamon Press, 11(2):225–232.

Demetracopoulos, A. C., L. Sehayek and H. Erodogan. 1986. "Modeling Leachate Production from Municipal Landfills," *Journal of the Environmental Engineering Division, ASCE,* 112(5):849–866.

DeWalle, F. B. and E. S. K. Chian. 1974. "Removal of Organic Matter by Activated Carbon," *Journal of the Environmental Engineering Division, ASCE,* 100(EE5): 1089–1104.

Emcon Associates. 1974. "Sonoma County Solid Waste Stabilization Study," EPA-SW 530 65D, Washington, D. C.: U.S. Environmental Protection Agency.

Fenn, D. G., K. J. Hanley and T. V. DeGeare. 1975. "Use of the Water Balance Method for Predicting Leachate Generation from Solid Waste Disposal Sites," EPA/530/SW-168, Cincinnati, Ohio: U.S. Environmental Protection Agency.

Fuller, W. H. 1978. "Investigation of Landfill Leachate Pollutant Attenuation by Soils," EPA-600/2-78-158, Cincinnati, Ohio: U.S. Environmental Protection Agency.

Fungaroli, A. A. 1971. "Pollution of Subsurface Water by Sanitary Landfills," Report SW-12 RG-71, Vol. I–III (NTIS No. PB-209 000), U.S. Environmental Protection Agency.

Fungaroli, A. A. and R. L. Steiner. 1971. "Laboratory Study of the Behavior of a Sanitary Landfill," *Journal Water Pollution Control Federation,* 43(2):252.

Fungaroli, A. A. and R. L. Steiner. 1979. "Investigation of Sanitary Landfill Behavior," Vol. I, Final Report, EPA-600/2-79-053a, Cincinnati, Ohio: U.S. Environmental Protection Agency, pp. 331.

Hentrich, R. L., J. F. Schwartzbaugh and J. A. Thomas. 1979. "Influence of MSW Processing on Gas and Leachate Production," *Municipal Solid Waste Land Disposal, Proceedings of the Fifth Annual Research Symposium,* EPA-600/9-79-023a.

Keenan, J. D., R. L. Steiner and A. A. Fungaroli. 1983. "Chemical-Physical Leachate Treatment," *Journal of the Environmental Engineering Division, ASCE,* 109(EE6): 1371–1384.

Kelly, W. E. 1976. "Groundwater Pollution Near a Landfill," *Journal of the Environmental Engineering Division, ASCE,* 102(EE6):1189–1199.

Kemper, J. M. and R. B. Smith. 1981. "Leachate Production by Landfilled Processed Municipal Wastes," in *Land Disposal of Municipal Solid Waste, Proceedings of the Seventh Annual Research Symposium,* EPA-600/9-81-002a, Cincinnati, Ohio: U.S. Environmental Protection Agency, pp. 18–36.

Leckie, J. G., J. G. Pacey and C. Halvadakis. 1979. "Landfill Management with Moisture Control," *Journal of the Environmental Engineering Division, ASCE,* Vol. 105, No. EE2:337–355.

Levine, A. D. and L. R. Rear. 1989. "Evaluation of Leachate Monitoring Data from Co-Disposal, Hazardous, and Sanitary Waste Disposal Facilities," *Proceedings of the 43rd Industrial Waste Conference,* Purdue University, West Lafayette, Indiana: pp. 173–182.

Liskowitz, J. W., P. C. Chan and R. B. Trattner. 1976. "Evaluation of Selected Sorbents for the Removal of Contaminants in Leachate from Industrial Sludges," in *Residual Management by Land Disposal, Proceedings of the Hazardous Waste Research Symposium,* EPA 600/9-76-015, Cincinnati, Ohio: U.S. Environmental Protection Agency, pp. 162–176.

Lu, J. C. S., B. Eichenberger and R. J. Stearns. 1984. "Production and Management of Leachate from Municipal Landfills: Summary and Assessment," EPA-602/2-84-092, Municipal Environmental Laboratory, Cincinnati, Ohio: U.S. Environmental Protection Agency.

Lu, J. C. S., B. Eichenberger and R. J. Stearns. 1985. *Leachate from Municipal Landfills, Production and Management,* Park Ridge, New Jersey: Noyes Publ.

Lu, J. C. S., R. D. Morrison and R. J. Stearns. 1981. "Leachate Production and Management from Municipal Landfills: Summary and Assessment," in *Land Disposal of Municipal Solid Waste,* 7th Annual Res. Symp., EPA-600/9-81/002a, 1-17, D. W. Shultz, ed., Cincinnati, Ohio: U.S. Environmental Protection Agency, pp. 1–17.

Miller, D. W., F. A. DeLucas and T. L. Tessier. 1974. "Ground Water Contamination in the Northeast States," Report on Contact 68-01-0777, Office of Research and Development, Washington, D.C.: U.S. EPA.

Pfeffer, J. T. 1986. "Treatment of Leachate from Land Disposal Facilities," *Proceedings of the WASTE TECH 86 Conference: Preparing Now for Tomorrow's Needs*, Oct. 22–23, 1986, Chicago, Illinois.

Phelps, D. H. (undated). "Solid Waste Leaching Model," Department of Civil Engineering, University of British Columbia, Report Submitted to Environment Canada, Solid Waste Branch.

Pohland, F. G. 1975. "Sanitary Landfill Stabilization with Leachate Recycle and Residual Treatment." EPA-600/2-75-043, Cincinnati, Ohio: U.S. Environmental Protection Agency.

Pohland, F. G., D. E. Shank, R. E. Benson and H. H. Timmerman. 1979. "Pilot-Scale Investigations of Accelerated Landfill Stabilization with Leachate Recycle," *Municipal Solid Waste Land Disposal Proceedings of the Fifth Annual Research Symposium*, EPA-600/9-79-023a.

Pohland, F. G., M. Stratakis, W. H. Cross, S. F. Tyahla. 1990. "Controlled Landfill Management of Municipal Solid and Hazardous Wastes," Proceedings of 1990 WPCF *National Specialty Conference on Water Quality Management of Landfills*, Chicago, Illinois, July 15–18, pp. 3–16, 3–32.

Qasim, S. R. 1965. "Chemical Characteristics of Seepage Water from Simulated Landfills," Ph.D. Dissertation, West Virginia University, Morgantown, W. Va.

Qasim, S. R. and J. C. Burchinal. 1970a. "Leaching from Simulated Landfills," *Journal of the Water Pollution Control Federation*, 42(3):371–379.

Qasim, S. R. and J. C. Burchinal. 1970b. "Leaching of Pollutants From Refuse Beds," *Journal of the Sanitary Engineering Division, ASCE*, 96(SA1):49–58.

Raveh, A. and Y. Avnimelech. 1979. "Leaching of Pollutants from Sanitary Landfill Models," *Journal Water Pollution Control Federation*, 51(11):2705–2716.

Rovers, F. A. and G. J. Farquhar. 1973. "Infiltration and Landfill Behavior," *Journal of the Environmental Engineering Division, ASCE*, 99(EE5):671–690.

Savage, G. M. and G. J. Trezek. 1980. "Significance of Size Reduction in Solid Waste Management," EPA-600/2-80-115, Cincinnati, Ohio: U.S. Environmental Protection Agency, pp. 138.

Schroeder, P. R., J. M. Morgan, T. M. Walski and A. C. Gibson. 1984. "The Hydrologic Evaluation of Landfill Performance (HELP), Model," Technical Resource Document, U.S. EPA PB-85-100840, and PB-85-100832, Cincinnati, Ohio.

Stone, R. 1974. "Disposal of Sewage Sludge into a Sanitary Landfill," EPA-SW-71d, Cincinnati, Ohio: U.S. Environmental Protection Agency.

Straub, W. A. and D. R. Lynch. 1982a. "Model of Landfill Leaching—Moisture Flow and Inorganic Strength," *Journal of the Environmental Engineering Division, ASCE*, 108(EE2):231–250.

Straub, W. A. and D. R. Lynch. 1982b. "Model of Landfill Leaching—Organic Strength," *Journal of the Environmental Engineering Division, ASCE*, 108(EE2): 251–268.

Tchobanoglous, G., H. Theisen and S. Vigil. 1993. "Integrated Solid Waste Management: Engineering Principles and Management Issues," New York: McGraw-Hill, Inc.

U.S. Environmental Protection Agency. 1988. "Summary of Data on Municipal Solid

Waste Landfill Leachate Characteristics Criteria for Municipal Solid Waste Landfills (40 CRF Part 258), EPA/530-SW-88-038, Office of Solid Wastes, Washington, D.C.

U.S. Environmental Protection Agency. 1993. "Standards for the Use or Disposal of Sewage Sludge," 40 CFR 503, February 19.

Walsh, J. J. and R. N. Kinman. 1979. "Leachate and Gas Production under Controlled Moisture Conditions," *Municipal Solid Waste Land Disposal, Proceedings of the Fifth Annual Research Symposium,* EPA-600/9-79-023a.

Walker, W. H. 1969. "Illinois Groundwater Pollution," *Journal American Water Works Association,* 61(1):31–40.

Wigh, R. J. 1979. "Boone County Field Site Interim Report," U.S. Environmental Protection Agency, EPA-600/2-79-058.

Wigh, R. J. and D. R. Brunner. 1979. "Leachate Production from Landfilled Municipal Waste: Boone County Field Site," *Municipal Solid Wastes: Land Disposal Proceedings of the Fifth Annual Research Symposium,* EPA-600/9-79-023a.

CHAPTER 7

Leachate Attenuation

7.1 INTRODUCTION

IN the past, most landfills were designed and built with the idea that the leachate would be attenuated (purified) by the natural soil beneath the landfill; thus groundwater contamination would not arise. Most soils do have attenuation properties, and are capable of purifying leachate to a certain degree. However, in recent years, studies have shown that even small landfills can adversely impact the groundwater quality if sites are not properly selected and landfills are not properly designed (Bagchi 1990; Kelly 1976). In this chapter, basic information on the migration of contaminants and attenuation processes of soils is briefly presented.

7.2 ATTENUATION PROCESS

Attenuation is a physical, chemical, and/or biological reaction or transformation that causes a temporary or permanent decrease in the concentrations of many contaminants of waste in a fixed time and distance travelled. Natural soils present a complex and dynamic system in which physical, chemical and biological reactions continually interact. Soils are a heterogeneous, polydispersed system of solid, liquid and gaseous components in various proportions. Soils are also very porous and chemically solvent bodies. They consist of (a) inert chemical compounds, (b) difficult, and easily soluble substances, (c) soluble salts, (d) complex insoluble compounds, and (e) a wide variety of organic compounds and organisms.

Soils present a suitable medium in which a series of complex biological activities occur simultaneously. The rate of interaction and the dominance of one reaction over the other are controlled by specific soil constituents. The constituents and their levels vary with parent material, time, climate, topography, and vegetation. Soil properties most useful in predicting the

151

mobility of waste constituents are: (a) texture (clay content) and particle size distribution, (b) content of hydrous oxides (Fe, Mn, and Al), (c) type and content of organic matter, (d) cation exchange capacity, and (e) soil pH.

The attenuation of leachate from a landfill occurs in two stages: (1) flow through the unsaturated zone, and (2) flow through the groundwater aquifer.

7.3 ATTENUATION MECHANISMS

Natural attenuation mechanisms may be categorized as physical, chemical and biological. The major attenuation mechanisms falling into these three categories are summarized below. Brief discussions on each of these mechanisms is presented in this section.

(1) Physical: (a) Filtration, (b) Diffusion and dispersion, (c) Dilution, and (d) sorption

(2) Chemical: (a) Precipitation/dissolution, (b) Adsorption/desorption (sorption), (c) Complexation, (d) Ion exchange, and (e) Redox reaction

(3) Microbiological: (a) Aerobic, and (b) Anaerobic

7.3.1 PHYSICAL MECHANISMS

Physical mechanisms are the physical forces in processes that bring about attenuation, or a reduction in concentration levels. The important physical mechanisms are briefly presented below.

7.3.1.1 Filtration

Leachate contains suspended particles that have a wide range of sizes. Filtration is applicable to all ranges of suspended and colloidal particles. The removal is mainly by straining action, although other mechanisms such as impaction, interception and orthokinetic flocculation also attribute to removal by filtration (Devinny et al. 1990). As particles accumulate in the pores, the permeability of the soil will decrease. It is physically possible to reduce the migration of leachate in the soil strata due to a reduction in permeability. The extent of attenuation achieved through filtration is difficult to estimate. The grain size and permeability of several soils is provided in Table 7.1.

TABLE 7.1. Properties of Soils and Their Suitability for Sludge Landfilling.

Soil Type	Grain Size (mm)	Permeability (cm/s)	CEC (meq/100 g)	Suitability for Landfilling
Clay	0.002 and less	10^{-8}–0^{-6}	Over 20	Excellent. Impermeable liner may not be needed
Silt loam	0.002–.05	10^{-6}–10^{-3}	12–20	Fair. Impermeable liner may be needed
Sandy soils	0.05–0.25	10^{-3}–10^{-1}	1–10	Poor. Impermeable liner needed for protection of groundwater

Source: Adapted in part from U.S. EPA (1978) and Qasim (1994).

7.3.1.2 Diffusion and Dispersion

Diffusion and dispersion are two mechanisms by which leachate is diluted by the aquifer. Molecular diffusion is caused by the concentration gradient of contaminants. As a result, a constituent moves from a region of high concentration to a region of low concentration. Mang et al. (1978) suggested that in the case of very low concentration of leachate flow rates, the diffusion of soluble species in the soil solution may be a significant migration mechanism.

Hydrodynamic dispersion is the result of variations in pore velocities within the soil. It is effective in attenuating the maximum constituent concentration rather than the total quantity of the constituent in a pulse or slug of leachate. Dispersion can occur in both longitudinal and transverse directions. The relative importance of diffusion and dispersion has been extensively studied by Perkins and Johnston (1963).

7.3.1.3 Dilution

Dilution reduces the concentration of leachate constituents due to mixing with groundwater. The ratio of contaminant dilution is proportional to the solution flux of both leachate and groundwater. The finer the soil texture, slower is the flow, and therefore, slower is the rate of dilution. Chloride, nitrate, hardness and sulfate found in municipal landfill leachate are not attenuated by soil. These constituents are attenuated only by dilution (Bagchi 1990).

7.3.1.4 Physical Sorption

Physical sorption is a function of van der Waals forces, and hydrodynamic and electrokinetic properties of soil particles. Only a small portion of the reaction of trace contaminants in soil/water solutions can be defined as physical adsorption. Bacteria and virus removal however, is an important physical adsorption mechanism (Gilbert 1976).

7.3.2 CHEMICAL MECHANISMS

Chemical mechanisms involve reactions that are basically chemical in nature. These mechanisms are briefly presented below.

7.3.2.1 Precipitation/Dissolution

Precipitation and dissolution are important reactions that control concentration levels and limit the total amount of contaminants in leachate when leaching through soils. The contaminant levels are usually governed by the solubilities of the solid. In particular, precipitation/dissolution reactions are important for migration of trace metals. The attenuation effects on metals is greatly controlled by the pH of the system. This topic is discussed in later sections.

7.3.2.2 Adsorption/Desorption

Adsorption is a process by which the molecules adhere to the surface of individual clay particles. Desorption is the opposite of adsorption, in which molecules leave the surface. Both processes are dependent upon the pH of the environment and the nature of the soil and waste contaminants. It is often the most common mechanism associated with the attenuation of trace contaminants. Adsorption will also cause a decrease in the total dissolved solids in leachate. The adsorption capacity of a soil is determined experimentally. Clay minerals, hydrated aluminum, iron and manganese oxides, and organics adsorb constituents. Because of pH-dependent charge characteristics, soils may exhibit sorption, and cation exchange simultaneously.

7.3.2.3 Complexation

Complexation or chelation is the formation of inorganic-organic complexes. The extent of leachate attenuation by complexation in the soil environment is not known, and is difficult to predict.

7.3.2.4 Ion Exchange

Clays have the property to exchange ions of one type with ions of another type. The total capacity of soils to exchange cations is called the *cation exchange capacity (CEC)*. The *CEC* of any particular soil is affected by the kind and quantity of clay mineral, organic content, and by the pH of the soil. The *CEC* of various soils is given in Table 7.1. In general, the silicate secondary minerals in soils hold a permanent negative charge. Therefore, the cation exchange property arises from the need to balance the negative charge of clay to maintain neutrality. To accomplish this, the positive ions in the soil solution become associated with the negative charge in the exchange complex (Lu et. al. 1985). These ions are mobile and readily exchange with other cations in the soil solution to maintain chemical equilibrium. The exchange capacity of a soil system generally depends upon (a) particle size, (b) organic content, and (c) pH. Soils containing smaller grains offer larger surface area and larger available exchange sites. Organic contents improve the exchange capacity, and the cation exchange capacity increases with increasing soil pH.

In nature, removal mechanisms of trace metals by ion exchange is not significant because other cations (calcium, magnesium, sodium and potassium), being in higher concentration in the leachates, utilize most of the active sites. The removal of trace metals by soils occurs simultaneously by adsorption, complexation and ion exchange; therefore these mechanisms are generally grouped together.

7.3.2.5 Redox Reaction

Redox reactions are oxidation and reduction reactions that greatly affect the solubilities of the contaminants. Iron and manganese in the oxidized state are less soluble. Attenuation of other trace metals in a reducing environment and in the presence of sufficient sulfide is more favorable due to lowered solubilities.

7.3.3 MICROBIOLOGICAL MECHANISMS

Biological decomposition of the organic component of leachate takes place in the subsoil structure. The microbial activity may be aerobic or anaerobic depending upon the availability of molecular oxygen. The biochemical reactions are complex. Under aerobic conditions the carbonaceous organic matter, ammonia, sulfide, phosphorus, iron and manganese are converted to carbon dioxide, nitrate, sulfate, phosphate and oxidized states of iron and manganese, respectively. Under anaerobic conditions, the carbonaceous matter is decomposed to produce organic acids, carbon dioxide, methane, and many other complex organic compounds. Denitrification and reduction of metals are other biochemical reactions of anaerobic activity. In general, the microbiological activity causes immobilization by conversion of organics and inorganics into cellular mass, and by precipitation of inorganics. It may also cause mobilization of organics by solubilization and conversion into smaller fragments, and solubilization of metals by reduction reactions and release under acidic conditions (carbonic and other organic acids).

7.3.4 MIGRATION TRENDS OF CONTAMINANTS

The migration trends of contaminants from sanitary landfills depend upon: (1) the characteristics of the soil, (2) characteristics of the leachate, and (3) environmental conditions and activity in the fill. The impact of major environmental factors and their effects upon various chemical constituents are described below.

7.3.4.1 pH and Redox Potential

Landfill leachate is generally acidic because of the accumulation of organic acids during the early life of the fill. Some neutralization may occur due to the dissolution of calcium carbonate and other minerals in the soil column. Many reactions in leachate are governed by the redox potential and pH. Among these are solubilization or precipitation of iron, manganese, other metals, sulfur and phosphorus, conversion of nitrogen, and other reactions.

7.3.4.2 Organic Matter

The migration of organic matter in the soil/water system is greatly influenced by microbiological activity, surface sorption, and chelation. Microbiological decomposition of organic matter in leachate and soil is a significant attenuating mechanism.

7.3.4.3 Alkalinity

Alkalinity in leachate is due to carbonates, bicarbonates, silicates, borates, ammonia, organic bases, sulfides and phosphates (Lu et al. 1985). Alkalinity in the soil is affected mainly by dissolution and precipitation of metal carbonates.

7.3.4.4 Major Ions

Major ions in leachate are sodium, potassium, calcium, magnesium, chloride, and sulfate. The attenuation of these ions depends upon solubilities and ion exchange. Dilution within the aquifer is also a major cause of concentration reduction of these ions.

7.3.4.5 Nutrients

Nitrogen and phosphorus are macronutrients. Nitrogen may exist as organic, ammonia, nitrite, or nitrate nitrogen. The transformation of nitrogen is dependent upon microorganisms, pH, and redox potential. Microorganisms play an important role in these conversions and attenuations. Other mechanisms are adsorption, ion exchange, and complexation. Nitrate ions are relatively mobile and are not retained by the ion exchange process.

Phosphorus compounds in the soil/water environment undergo complex physical, chemical and microbiological transformations. The attenuation mechanisms for phosphorus compounds are microbial uptake, precipitation, complexation, solubilization and sorption. The solubility of phosphate in leachate depends upon pH and alkalinity.

7.3.4.6 Trace Metals

Movement of trace metals in the soil/water environment is extremely complex. Major mechanisms that influence the mobility of trace metals are: (1) precipitation/solubilization, (2) sorption, (3) ion exchange, (4) complexation/chelation, and (5) dilution. Each metal behaves differently in the soil/water environment. The governing environmental factors that influence the mobility of metals are pH, redox potential, microbiological activity and soil chemistry. Table 7.2 (pages 160–171) describes the major attenuation mechanisms of various leachate constituents. Most metals attenuate well in clayey soils; non-metals are not attenuated well.

7.3.4.7 Chlorinated Hydrocarbons and Pesticides

Chlorinated hydrocarbons and pesticides are attenuated mainly by sorption. Other mechanisms are volatilization, microbiological degradation, hydrolysis, oxidation and dilution. The adsorption and attenuation of chlorinated hydrocarbons and pesticides increase with an increase in clay content.

7.3.4.8 Viruses

Virus survival in soil depends upon the pH, temperature, moisture content, nutrients and antagonism (Keswick and Gerba 1980). Viruses survive longer in soil than at the surface. They can travel longer distances in soil. The specific factors that control their travel distances are soil composition, pH, soluble organics, and leachate quantity. Yates et al. (1992) presented models of virus transport in unsaturated soil.

7.4 LEACHATE MIGRATION PREDICTION AND MODELING

7.4.1 MODEL DESCRIPTION

Many attempts have been made in the past to develop models that will predict the travel of pollutants in subsurface soils originating from natural attenuation landfills. These models approximately simulate the attenuation and transformation processes to evaluate the migration patterns of the contaminants. Bear et al. (1992) suggested that since real-world systems are very complex, there is a need for simplification. Simplification is achieved by introducing a set of assumptions which approximate the real situations. Examples are isotropy, homogeneity, properties of fluids, flow regime (laminar or nonlaminar), solid flux, boundaries, mode of flow (one-, two-, or three-dimensional), and others. Lu et al. (1985) suggested that leachate migration models may be categorized as: (1) descriptive models, (2) physical models, (3) analog models, and (4) mathematical models.

Descriptive models generally utilize a qualitative judgement approach, rather than well defined quantitative data. These models are subjective, site-specific, and have limitations. Physical models are generally scaled-down versions of waste sites and attenuation information, developed from experimental data. Analog models utilize transformation of a given physical system into another system that can be measured more easily. Electrical analog models have found application in groundwater flow modeling. Mathematical models employ a set of concise mathematical equations to describe the relationship between various system parameters and their input and output variables. These equations can be solved using analytical or numerical techniques.

Lu et al. (1985) provided several distinctions between models, depending upon the method of analysis defined by the model and the particular approach used to solve the model. Common model distinctions are provided in Table 7.3.

The most promising and generally utilized models for leachate transport prediction and attenuation are the conceptual mathematical models. These models are generally two or three dimensional and apply to

- saturated transport
- unsaturated transport
- saturated and unsaturated transport
- analytical transport

Saturated models describe the transport into the groundwater zone. This is a special case if the landfill extends into a groundwater aquifer. The unsaturated transport model deals with the movement of leachate above the groundwater table. These models can be used to predict the quantity and quality of leachate reaching the groundwater. The analytical models are generally restricted to identification and quantification of waste-soil interactions with leachate studies.

The conceptual mathematical models utilize a set of mathematical equations defining input and output functions, variables and relationships. Among these may be changes in concentration due to adsorption, ion exchange, diffusion/dispersion, convection, dilution, permeability, production or decay, and others. After simplification, the equations are reduced to a set of non-linear, second-order partial differential equations defining fluid and constituent movement and transport behavior. Numerous conceptual mathematical models are presently available that differ from each other due to simplifying assumptions, basic derivation of equations, stipulated boundary conditions and methods of solution.

TABLE 7.2. Major Attenuation Mechanisms of Various Constituents of Landfill Leachates.

Landfill Leachate Constituent	Major Attenuation Mechanism(s)	Factors Affecting Attenuation	Description	Mobility in Clayey Environment
CATIONS				
Aluminum	Precipitation	pH	Aluminum readily forms insoluble oxides, hydroxides, silicates above pH 7. Below pH 7, aluminum may solubilize.	Low
Ammonium	Cation exchange, nitrification or biological uptake	Soil texture pH Oxygen supply moisture content temperature	Nitrification is maximum at a pH around 8.5; adequate ammonia and oxygen, moderate moisture content and temperature around 30°C. Coarser soils increase transformation.	Moderate
Arsenic	Precipitation, and adsorption	pH, type of clay	In aerobic environment arsenic reacts with iron, aluminum, calcium and other metals to form slightly soluble arsenate; maximum removal of arsenate occurs when the pH is between 4 and 6. Arsenite removal is maximum in the pH range of 3 and 9. Montmorillonite clay adsorbs twice as much arsenic as kaolinite clay.	Moderate

TABLE 7.2. (continued).

Landfill Leachate Constituent	Major Attenuation Mechanism(s)	Factors Affecting Attenuation	Description	Mobility in Clayey Environment
Barium	Adsorption, ion exchange, and precipitation	Cation exchange capacity (CEC), pH and percent clay in soil	As the CEC increases, adsorption of barium also increases; alkaline conditions and free lime favor attenuation by ion exchange and chemical precipitation. Attenuation is also favored by a high clay percentage and the presence of other colloidal material.	Low
Beryllium	Precipitation, and cation exchange	CEC and type of clay	Mobile in both high and low pH due to hydrolysis; highly attenuated in soils containing montmorillonite and illite.	Low
Boron	Adsorption, precipitation or coprecipitation	Influenced by the activities of Al and Fe (III)	Adsorbed as borate on inorganic surfaces and precipitated or coprecipitated with hydrous iron or aluminum oxides.	High
Cadmium	Precipitation, adsorption	pH, redox potential	Undergoes hydrolysis at normal pH; attenuation of cadmium is very significant, between pH 6 and 8, precipitation occurs at near neutral environment; anions such as phosphate, sulfide, carbonate effectively attenuate cadmium by precipitation.	Moderate

(continued)

161

TABLE 7.2. (continued).

Landfill Leachate Constituent	Major Attenuation Mechanism(s)	Factors Affecting Attenuation	Description	Mobility in Clayey Environment
CATIONS (continued)				
Calcium	Precipitation, and cation exchange	pH, CEC, and type of clay	Readily forms carbonate that precipitates in alkaline conditions. Since calcium is seldom adsorbed but often eluted, hardness of groundwater beneath a landfill increases. Montmorillonite elutes calcium more significantly than illite or kaolinite.	High
Chromium	Precipitation, cation exchange, and adsorption	Redox potential, form of chromium pH, type of clay, and soil texture	The importance of each attenuation mechanism depends on the form of chromium (hexavalent or trivalent). Redox potential has a marked effect on Cr attenuation. The concentration of Cr in the soil is reduced by adsorption on organic matter, clay minerals, and hydrous oxides of iron, manganese, aluminum, and precipitation as an oxide. Montmorillonite clay is more effective than kaolinite clay in Cr attenuation.	Low for trivalent chromium. High for hexavalent chromium.

TABLE 7.2. (continued).

Landfill Leachate Constituent	Major Attenuation Mechanism(s)	Factors Affecting Attenuation	Description	Mobility in Clayey Environment
Copper	Adsorption, ion exchange, and chemical precipitation	Type of clay and pH	Soil pH is the most important factor controlling removal of copper with a given adsorbent. Some copper compounds dissolve under acidic conditions. In the pH range of 5–6, precipitation of copper compounds can occur when copper concentrations are high. Complexing with organic matter is high. Soil materials favoring attenuation of copper include colloidal matter, free lime, hydrous oxides of manganese and iron; high clay content, and organic matter.	Low

(continued)

163

TABLE 7.2. (continued).

Landfill Leachate Constituent	Major Attenuation Mechanism(s)	Factors Affecting Attenuation	Description	Mobility in Clayey Environment
CATIONS *(continued)*				
Iron	Precipitation, cation exchange, adsorption, and biological uptake	CEC, pH, and Redox potential	Iron compounds are attenuated moderately in soil. The form of iron affects attenuation [Fe (II) or Fe (III)]. Iron attenuation increases with increase in CEC. Mobility is increased when ferric iron is converted to ferrous iron due to reducing conditions created by anaerobic growth.	Low for trivalent iron and high for divalent iron. Since it is difficult to assess which species is present in leachate, the mobility of iron is assumed to be moderate.

TABLE 7.2. (continued).

Landfill Leachate Constituent	Major Attenuation Mechanism(s)	Factors Affecting Attenuation	Description	Mobility in Clayey Environment
Lead	Adsorption, cation exchange, and precipitation	pH, and type of clay	Attenuation increases as the pH rises above 5. Montmorillonite has a higher removal efficiency than kaolinite. Lead forms poorly soluble precipitates with sulfate, carbonate, phosphate and sulfide anions. Soil materials favoring attenuation of lead includes organic matter, clays and free lime. Lead has a strong affinity to organic matter that results in complexion and immobilization. It is attenuated more than many other divalent heavy metals.	Low
Magnesium	Cation exchange, and precipitation	pH	When pH is between 7 and 14, magnesium forms carbonate and hydroxide precipitate. It is attenuated moderately in clayey soil.	Moderate

(continued)

TABLE 7.2. (continued).

Landfill Leachate Constituent	Major Attenuation Mechanism(s)	Factors Affecting Attenuation	Description	Mobility in Clayey Environment
CATIONS *(continued)*				
Manganese	Precipitation, and cation exchange	Redox potential, valence of manganese, and type of clay	Manganese normally exists as insoluble oxides. Under reducing conditions Mn (II) is formed. This increases solubility. Adsorption of manganese is highest in bentonite (montmorillonite), intermediate in illite, and lowest in kaolinite. Soil material favoring manganese attenuation are clays, organic matter, hydrous metal oxides and free lime. Alkaline condition and an abundance of anions such as sulfide and carbonate will improve retention.	High
Mercury	Adsorption, precipitation, and redox reactions	pH, type of clay, and colloidal matter in soil	Attenuation mechanisms usually cause the volatilization of mercury. Bacteria convert inorganic mercury into toxic mono- or dimethyl mercury. Clayey soils attenuate mercury by adsorption with iron oxide, organic matter, and clays. Maximum attenuation under alkaline conditions. Attenuation is improved by colloidal matter in clays, and iron oxides.	High

TABLE 7.2. (continued).

Landfill Leachate Constituent	Major Attenuation Mechanism(s)	Factors Affecting Attenuation	Description	Mobility in Clayey Environment
Nickel	Sorption, precipitation	Surface area, CEC, clay content, and pH	Removed by hydrous metal oxide precipitates. Factors favoring retention include alkaline conditions, high concentrations of hydrous metal oxides, and free lime.	Moderate
Potassium	Precipitation and cation exchange	pH	Potassium is well attenuated in clayey soil. Attenuation is maximum under neutral to alkaline conditions.	Moderate
Selenium	Adsorption, and anion exchange	pH, and type of clay	Present in soils as an inorganic anion, and associated with iron, sodium or calcium. Microorganisms oxidize or reduce it repeatedly. Selenium removal by montmorillonite is significantly higher than that by kaolinite. Maximum removal is in the pH range between 2 and 4.	Moderate
Sodium	Cation exchange	CEC	May be totally attenuated but since it is a monovalent ion a low concentration of sodium could pass through the soil.	Low to High

(continued)

TABLE 7.2. (continued).

Landfill Leachate Constituent	Major Attenuation Mechanism(s)	Factors Affecting Attenuation	Description	Mobility in Clayey Environment
CATIONS (continued)				
Zinc	Adsorption, cation exchange, and precipitation	pH, clay type, CEC, and organic matter	Favorably attenuated under alkaline conditions by clays, organics, hydrous metal oxides, and free lime. Attenuation is maximum in the pH range of 6–8. It also precipitates with a variety of anions including carbonate, sulfate, silicate and phosphate. pH and organic matter also affect the attenuation of zinc.	Low
ANIONS				
Chloride	Dilution	Soil pH	Chloride is not attenuated by any soil type. Dilution is the only attenuation mechanism.	High
Cyanide	Adsorption	Soil pH	Adsorption is the only attenuation mechanism. Cyanide is not strongly retained in soils. Adsorption of cyanide is dependent on the pH of the soil.	High
Fluoride	Anion exchange	Soil pH	Acidic soils adsorb fluoride more readily than alkaline soils. Solubility of fluoride increases in both acidic and alkaline soils.	High

TABLE 7.2. (continued).

Landfill Leachate Constituent	Major Attenuation Mechanism(s)	Factors Affecting Attenuation	Description	Mobility in Clayey Environment
Nitrate	Biological uptake, reduction	pH, redox potential, carbon source, microorganism	Biological denitrification or reduction to free nitrogen requires anaerobic condition and a carbon source	High
Phosphate	Solubilization, sorption, and biological effects	pH, alkalinity, and type of clay	The major species in leachates is orthophosphate. High adsorption by clays is favored by lower pH. Maximum sorption by montmorillonite occurs at pH 5–6, and by koalinite at a pH near 3. Solubilization of phosphates is favored at higher alkalinity.	High
Silica	Precipitation	pH	Silica readily precipitates in silicate mineral phases. Mobility increases under alkaline conditions.	Moderate
Sulfate	Anion exchange and dilution	pH and redox potential	Sulfate is relatively weakly held. Losses due to leaching are proportional to the amount of water passing through the soil—microorganism reactive environment. Sulfide loss as hydrogen sulfide gas may occur in the absence of iron.	High

(continued)

TABLE 7.2. (continued).

Landfill Leachate Constituent	Major Attenuation Mechanism(s)	Factors Affecting Attenuation	Description	Mobility in Clayey Environment
ORGANICS				
Chlorinated Hydrocarbons	Sorption, volatilization, microbial degradation, chemical hydrolysis and oxidation	pH, type of soil, microorganism	These tend to be strongly sorbed by soils. The major adsorbents are clay minerals, iron and manganese, hydrated oxides, and organic material. The rate of adsorption is rapid. Sorption is generally the dominant mechanism.	Low
COD	Biological uptake, and filtration	Availability of oxygen, pH, supply of nutrients, and soil texture	COD is effectively attenuated by biological uptake, and filtration of organisms, other mechanisms of attenuation are adsorption and ion exchange. Fine grain soil favors COD attenuation due to larger surface area.	Moderate
Pesticides	Adsorption/desorption, microbial decomposition, and volatilization	pH, soil moisture, soil properties, organic matter content, and redox potential	Pesticides may be cationic and anionic. Cationic pesticides are strongly retained on the exchange complex. Attenuation depends upon organic content, pH, moisture, and microbiological activity.	Moderate

TABLE 7.2. (continued).

Landfill Leachate Constituent	Major Attenuation Mechanism(s)	Factors Affecting Attenuation	Description	Mobility in Clayey Environment
Volatile Organic Compounds (VOC)	Biological uptake, sorption, dilution	Temperature, permeability, pH, fluid flow, microbiological activity	VOCs volatilize at normal temperature and pressure. Biodegradation of organic matter and the rate of metabolism is the main factor affecting attenuation.	Moderate
MICROORGANISMS				
Viruses	Unknown	Soil moisture content, pH temperature, nutrient availability, and water flow	Virus removal occurs in clayey soils at low pH, and in the presence of cations. Inactivation of viruses near the soil surface is much more rapid than when it penetrates into the soil. Transportation to long distances within the soil is possible.	Low

Source: Partly adapted from Roy F. Weston, Inc. (1978), Lu et al. (1985), Bagchi (1990), and Yates et al. (1992).

TABLE 7.3. **Model Distinctions.**

Distinction	Details
Empirical or Conceptual Models	Empirical models are based on observations and/or experimentation. Conceptual models use equations of mass, energy, and momentum.
Deterministic or Stochastic Models	Deterministic models utilize input variables and system parameters that have fixed mathematical or logical relationships with each other. These relationships completely define the system. Two categories of stochastic models are: (a) where the system parameters and input variables are categorized by assumed probability distribution (normal, log-normal, etc.), and (b) where system parameters or input variables are uncertain either because of unreliable data, or due to measurement errors.
Static or Dynamic Model	Utilize steady-state, dynamic conditions, and input and output variable changes with time.
Special Dimensionality Model	These models utilize the dimensionality as one, two, or three dimensions. Three dimensional models utilize more representative characteristics of the actual setting. There are also complex models.

Source: Adapted in part from Roy F. Weston, Inc. (1978), and Lu et al. (1985).

7.4.2 MODEL APPLICATION

The application of these models requires an understanding of the models, site conditions, estimation of the coefficients, calibration, and solution. Many uncertainties are associated with the modeling of a particular coefficient, initial conditions, location of domain boundaries and prevailing conditions, calibration, heterogeneity of strata, and ability of models to handle the specific situations.

7.4.2.1 Simplified Approach

Over the past 30 years efforts have been made to determine the impact of sanitary landfills and other waste impoundments upon groundwater quality. LeGrand (1969) proposed a *point count* system based on nomographs. These points are used to determine the pollution impact of a landfill upon the groundwater. Nomographs utilize variables such as soil texture, permeability, depth of the water table, adsorptive capacity, water table gradient, distance from the well, and thickness of the impervious layer. The points under each category are added up to give a final number that is used to evaluate the potential for pollution.

Hagerty et al. (1973) presented the *site ranking* system proposed by Booz Allen Applied Research, Inc. (1972) for hazardous waste sites. The method utilizes simplified empirical equations that are used to calculate points for site specific factors. A total of 10 such factors are used. These factors are

(1) Infiltration potential
(2) Bottom leakage potential
(3) Filtration capacity
(4) Adsorptive capacity
(5) Organic content of groundwater
(6) Buffering capacity of groundwater
(7) Travel distance
(8) Groundwater velocity
(9) Prevailing wind direction
(10) Population factor

The values calculated from these equations, using specific site conditions, are added together. The final number will provide the pollution potential and environmental impact of the landfill.

7.4.2.2 Complex Models and Applications

Lu et al. (1984) and Roy F. Weston, Inc. (1978) provided examples, equations, applications, case histories, and references on over 51 transport models that have been applied to landfill leachates and various waste impoundments. Walton (1989a) provided two microcomputer programs for simulating quasi-three dimensional groundwater flow and contaminant migration models using numerical and analytical techniques. Walton (1989b) also developed four analytical microcomputer programs for quick and easy simulation and graphing of uncomplicated two dimensional groundwater flow contaminant migration situations. Model simulation, equations, program listing and application are presented in detail in these sources.

U.S. EPA (1989) provided information on modeling subsurface contaminant transport and fate, and management considerations. Bear et al. (1992) compiled a review paper on the fundamentals of groundwater modeling and discussed analytical models, uncertainty, and model misuses. These researchers also provided sources of information dealing with groundwater modeling and fate of contaminants in the subsurface. For a large number of groundwater models, such information is available from the International Ground Water Modeling Center (IGWMC, Institute for Ground-Water Research and Education, Colorado School of Mines, Golden, Colorado 80401), which operates a clearinghouse service for information and software pertinent to groundwater modeling. Information databases have been developed to efficiently organize, update and access information on groundwater models for mainframe and microcomputers. The model annotation databases have been developed and maintained over the years with major support from US EPA's Robert S. Kerr Environmental Research Laboratory (RSKERL), Ada, Oklahoma.

The Center for Subsurface Modeling Support (CSMoS), located at the RSKERL (P.O. Box 1198, Ada, OK 74820), provides a groundwater and vadose zone modeling software and services to public agencies and private companies throughout the nation. CSMoS primarily manages and supports groundwater models and databases resulting from research at RSKERL. CSMoS integrates the expertise of individuals in all aspects of the environmental field in an effort to apply models to better understand and resolve subsurface problems. CSMoS is supported internally by RSKERL scientists and engineers, and externally by the IGWMC National Center for Ground Water Research and numerous groundwater modeling consultants from academia and the private consulting community.

The National Ground Water Information Center (NGWIC, 6375 Riverside Drive, Dublin, Ohio 43017) is an information gathering and dissemination business that performs customized literature searches on various groundwater related topics, and locates and retrieves copies of available documents. The center maintains its own on-line databases (Bear et al. 1992).

7.5 REFERENCES

Bagchi, A. 1990. "Design, Construction and Monitoring of Sanitary Landfills," New York: John Wiley and Sons.

Bear, J., M. S. Beljin, and R. R. Ross. 1992. "Fundamentals of Ground-Water Modeling," U.S. Environmental Protection Agency, EPA/540/S-92/005, Ada, OK: Robert S. Kerr Environmental Research Laboratory.

Booz Allen Applied Research, Inc. 1972. "Study of Hazardous Waste Materials, Hazardous Effects, and Disposal Methods," Report to Solid Waste Management Office, Cincinnati, Ohio: U.S. Environmental Protection Agency.

Devinny, J. S., L. G. Everett, J. C. S. Lu, and R. L. Stollar. 1990. "Subsurface Migration of Hazardous Wastes," New York: Van Nostrand Reinhold.

Gilbert, R. G. 1976. "Virus and Bacteria Removal from Wastewater by Land Treatment," *Applied and Environmental Microbiology,* Vol. 32, pp. 333–338.

Hagerty, D. J., J. L. Pavani, and J. E. Heer. 1973. "Solid Wastes Management," New York: Van Nostrand Reinhold Company.

Kelly, W. E. 1976. "Groundwater Pollution Near a Landfill," *Journal of the Environmental Engineering Division, ASCE,* 102(EE6):1189–1199.

Keswick, B. H. and C. P. Gerba. 1980. "Virus in Groundwater," *Environmental Science and Technology,* Vol. 14, pp. 1290–1297.

LeGrand, H. E. 1969. "Systems for Evaluation of Contamination Potential of Some

Waste Disposal Sites," Training Course Manual Sanitary Landfill Principles, U.S. Department of Health, Education and Welfare, Public Health Service, Washington, D.C.

Lu, J. C. S., B. Eichenberger, R. J. Stearns, and I. Melnyk. 1984. "Prediction and Management of Leachate from Municipal Landfills: Summary and Assessment," EPA 600/2-84-092, Environmental Research Laboratory, Office of Research and Development, U.S. Environmental Protection Agency, Cincinnati, Ohio.

Lu, J. C. S., B. Eichenberger, and R. J. Stearns. 1985. "Leachate from Municipal Landfills: Production and Management," Park Ridge, New Jersey: Noyes Publications.

Mang, J. L., J. C. S. Lu, R. L. Lofy, and R. J. Stearns. 1978. "A Study of Leachate from Dredged Material in Upland Areas and/or in Production Uses," Waterways Experiment Station, U.S. Army Corps of Engineers, Vicksburg, Mississippi.

Perkins, T. K. and O. C. Johnston. 1963. "A Review of Diffusion and Dispersion in Porous Media," *Society of Petroleum Engineers Journal*, 3(1):70–84.

Qasim, S. R. 1994. "Wastewater Treatment Plants: Planning, Design and Operation," Lancaster, PA: Technomic Publishing Co., Inc.

Roy F. Weston, Inc. 1978. "Pollution Prediction Techniques for Waste Disposal Siting; A State-of-the-Art Assessment," U.S. Department of Commerce, *National Technical Information Service*, PB-283 572 U.S. Environmental Protection Agency, Office of Solid Wastes, Washington, D.C.

U.S. Environmental Protection Agency. 1978. "Process Manual, Municipal Sludge Landfills," EPA-625/1-78-010, SW 705, U.S. Environmental Research Information Center, Technology Transfer, Office of Solid Wastes, Washington, D.C.

U.S. Environmental Protection Agency. 1989. "Transport and Fate of Contaminants in the Subsurface," Seminar Publication, EPA/625/4-89-019, Center for Environmental Research Information, Cincinnati, OH, and Robert S. Kerr Environmental Research Laboratory, Ada, OK.

Walton, W. C. 1989a. "Numerical Groundwater Modeling, Flow and Contaminant Migration," Chelsea, MI: Lewis Publishers, Inc.

Walton, W. C. 1989b. "Analytical Groundwater Modeling, Flow and Contaminant Migration," Chelsea, MI: Lewis Publishers, Inc.

Yates, M. V., S. R. Yates, and Y. Ouyang. 1992. "A Model of Virus Transport in Unsaturated Soil," Project Summary, EPA/600/S2-91-062, U.S. Environmental Protection Agency, Robert S. Kerr Environmental Research Laboratory, Ada, OK.

CHAPTER 8

Leachate Containment

8.1 INTRODUCTION

IN modern sanitary landfill design, leachate collection systems are used to limit the migration of potential leachate and to protect the groundwater from contamination. A collection system is also needed to remove the leachate for treatment and disposal. Liners are installed along the bottom and on the sides of a landfill to reduce the migration of leachate to groundwater beneath the site, as well as laterally. The liner might be constructed of a compacted clay or mixed materials, a prefabricated synthetic material, or a combination of the two. Synthetic liners, although essentially impermeable under ideal conditions, often leak under field conditions. Therefore, synthetic liners are sometimes placed over clay liners for additional safety.

In this chapter a review of various types of impermeable liners, material, construction methods, performance, and methods of leachate collection and removal systems are presented.

8.2 TYPES OF IMPERVIOUS LINERS

Sanitary landfill liners must be constructed of materials that have appropriate chemical properties and strength, and a sufficient thickness to prevent failure from internal or external pressures. The liner must also rest on a foundation or base capable of providing support and resistance to settlement or buckling. Liners, in general, function in two ways: (1) they impede the flow of the pollutants and pollutant carriers, and (2) they adsorb or attenuate suspended or dissolved constituents. A liner with low permeability is needed to impede pollutants. The adsorptive or attenuative capacity of a liner depends on its chemical composition and its mass. Liners can be classified in a variety of ways. Among these are: construction methods,

TABLE 8.1. Classification of Liners for Waste Disposal Facilities.

Basis	Description
A. Construction methods	On-site construction: • raw materials brought to site and liner constructed on site • compacted soil • mixed on site or brought to site mixed • sprayed-on liner Prefabricated: • drop-in polymeric membrane liner Partially prefabricated: • panels brought to site and assembled on prepared base
B. Structural properties	• Rigid (some with structural strength) – soil – soil cement • Semirigid – asphalt concrete • Flexible (no structural strength) – polymeric membranes – sprayed-on membranes
C. Materials and method of application	• compacted soils and clays • admixes, e.g., asphalt concrete, soil cement • polymeric membranes, e.g., rubber and plastic sheeting • sprayed-on linings • soil sealants • chemisorptive liners

Source: Matrecon, Inc. (1980).

physical properties, permeability, composition, and type of service. These classifications are presented in Table 8.1. (Matrecon, Inc. 1980).

Most liners incorporate flow-control and filtration mechanisms, but to different degrees. Membrane liners (FMLs) are the most impermeable, but have little adsorptive capacity. Soils have a larger adsorptive capacity, but can be more permeable. However, greater thickness of the soil liner will have lower potential for movement of pollutants through it. Due to their availability, soils normally are considered as the first alternative for landfill liners. Synthetic liners use materials constructed or fabricated by man, and include soils and clays of low permeability, either available at the site or brought to the site and compacted with additives, to further reduce permeability and increase strength. Cheremisinoff et al. (1979), Lu et al. (1984), Haxo et al. (1985), Loehr (1987), Matrecon, Inc. (1980, 1988), U.S. Congress (1989), Bagchi (1990), Goldman et al. (1990), and Tchobanoglous et al. (1993) provided discussions on various types of liners and installation requirements. The liners are classified into six categories. These are:

(1) Natural Soil and Clay Systems

(2) Admixed Liners: (a) asphalt concrete, (i) soil cement, and (ii) soil asphalt

(3) Sprayed-on Linings: (a) air blown asphalt, (b) membranes of emulsified asphalt, (c) urethane modified asphalt, and (d) rubber and plastic latexes

(4) Soil Sealants

(5) Polymeric Membrane: (a) butyl rubber, (b) chlorinated polyethylene (CPE), (c) chlorosulfonated polyethylene (CSPE), (d) elasticized polyolefin (ELPO), (e) epichlorohydrin rubber (CO and ECO), (f) ethylene propylene diene mono rubber (EPDM), (g) neoprene, (h) polyethylene, (i) polyvinyl chloride (PVC), and (j) thermoplastic elastomer (TPE)

(6) Composite Liner

8.2.1 NATURAL SOIL AND CLAY SYSTEMS

Due to their availability, clayey soils should be considered as the first alternative for a waste confinement liner. The native material must be evaluated and should be used first. If the result of such analysis is negative, the soil from other sources must be evaluated for treatment, remolding and compaction to increase strength and reduce permeability. Bentonite is often used for subgrade cover in areas where compatibility or soil tests show that the proper application will lower the permeability to the desired level. Common application systems include spreading, mixing, and compacting. Bentonite is an extremely absorbant, porous clay, which holds liquid and becomes impermeable. In general, clay liners are more permeable to water than are synthetic liners. Engineered soils, however, are less permeable than uncompacted soils. The permeability of natural soil liners to organic chemicals is variable. It depends on the characteristics and concentration of the chemicals, degree of compaction, and other engineering properties of the soil. Compacted clay liners can adsorb much of the organic pollutants in leachate, however, little is known about adsorptive capacity for chemical solvents (Bingemer and Crutzen 1987). Soil liners can become desiccated by some solvents that are insoluble in water (e.g., xylene and carbon tetrachloride) causing water to migrate out of the soil. When desiccation occurs, the soil may shrink and channeling of the soil may form pathways through which liquids can flow. Daniels (1988) indicated that the majority of liquid flow through clay liners takes place through macro cracks created by desiccation or improper placement and compaction rather than permeation through the micropores of the clay liner (Figure 8.1). Clay liners are often placed and compacted in 15–20 cm lifts (6–8 inch). It is important to *scarify* the previous lift prior to placing the next lift. This eliminates cold joints from happening between the lifts. Cold joints are the leading path for in-plane flow of liquids through a clay liner system. Therefore, construction technique and quality assurance monitoring are very important factors in success or failure of a clay liner. Soils to be used as liners at waste disposal facilities must contain a relatively large proportion of fines, i.e., particle size less than 2 μm. For a broad range of soils an inverse relationship exists between soil permeability and proportion of fines. The minimum amount of clay-size particles required in soil to yield a good soil liner is 25–30 percent by weight.

Bagchi (1990), reported that the mechanical properties of clay depend upon several interacting factors such as mineral composition, percentage of amorphous material, adsorbed cation, distribution and shape of particle, pore fluid chemistry, soil fabric, and degree of saturation. Various physiochemical theories can be used to predict quantitatively the change in mechanical properties of soil due to change in any of these factors. Discussions on this subject may be found in many references (Acar and Seals 1984; Acar and Ghosh 1986; Matrecon, Inc. 1980, 1988; Bagchi 1990; Goldman et al. 1990). For sanitary landfills, the criteria for choosing a clay are primarily based on the recommended permeability achievable under field conditions. Bagchi (1990) noted that a clay that can obtain a low permeability (1 $\times$ 10^{-7} cm/s or less) when compacted to 90–95% of the

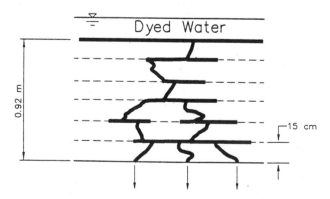

Figure 8.1 Permeation of dye tracer through clay liner constructed in six compacted lifts 15 cm each. Source: Daniels (1988).

TABLE 8.2. Properties of Soil Admixes Used as Liners.

	Asphalt Concrete[a]	Asphalt Concrete[b]	Soil Cement[c]	Soil Asphalt[d]
Thickness cm	5.6	6.1	11.4	10.2
Density, g/cm³	2.39	2.52	2.17	2.23
Permeability, cm/s	1.2×10^{-8}	3.3×10^{-9}	1.5×10^{-6}	1.7×10^{-3}
Compressive strength after 25 hrs, kN/m² (psi)	15,400 (2230)	16,000 (2328)	9,100 (1323)	1,300 (184)

[a]7 units asphalt and 100 units aggregate.
[b]9 units asphalt and 100 units aggregate.
[c]95 units soil, 5 units koalin clay, 10 units cement, and 8.8 units water.
[d]7 units liquid asphalt and 100 units aggregate.
Source: U.S. EPA (1983).

maximum Proctor's dry density should be chosen for landfill liner construction. Bagchi (1990), also noted that clay with a high liquid limit (LL) tends to develop more desiccation cracks, clay with a very low plasticity index (PI) or plastic limit (PL) is less workable, and a well graded soil is expected to develop low permeability when compacted properly. Therefore, the PI, LL, and some minimum requirements regarding grain size distribution should be specified. The recommended specification for a low permeability liner is: PI 10–15%; LL 25–30%; 40–50% fraction having particle size 0.074 mm or less; and 18–25% clay content (Bagchi 1990). A clayey soil liner can attenuate a pollutant by reducing the level of contamination and spreading the contamination over a much larger period.

8.2.2 ADMIXED LINERS

A number of admixed or formed-in-place liners have been used to impound water. Possible admixes include asphalt concrete, soil cement, and soil asphalt. Example properties of such liners are noted in Table 8.2. Reactions of the landfilled material with these mixtures should be evaluated so that only compatible wastes are used with these liners. A brief description of each type of admixed liner is given below (Haxo et al., 1985; Loehr, 1987; Matrecon, Inc. 1980, 1988).

8.2.2.1 Asphalt Concrete

Asphalt concretes are controlled hot mixtures of asphalt cement and high quality mineral aggregate, compacted into a uniform dense mass.

They are similar to highway paving asphalt concrete but have a higher percentage of mineral fillers, and a higher percentage of asphalt cement. The asphalt used is usually a hard grade. These harder grade asphalts are better suited as liners than softer paving asphalt. Asphalt concrete can be compacted to have a permeability coefficient less than 1×10^{-7} cm/s. It is resistant to erosion, growth of plants, and effects of weather extremes (temperature). Such asphalt concrete is stable on side slopes, resists slip and creep, and retains enough flexibility to conform to slight deformations of the subgrade, and avoids rupture from low level seismic activity. The hydraulic mix for asphalt concrete is designed with a properly graded aggregate, and high mineral filler and asphalt cement contents (The Asphalt Institute 1976). These liners may be placed with conventional paving equipment and compacted to the required thickness. Styron and Fry (1979) used 11 percent asphalt in a 5 cm asphalt concrete liner to obtain the necessary permeability. Haxo and White (1976) used nine percent asphalt concrete to test its durability. After one year of exposure to leachate from a simulated landfill, they determined that liner thickness greater than 9 cm may be necessary to contain the waste. Extra thickness may be necessary due to potential inhomogeneities in the admixed materials resulting from inadequate mixing or compaction. The liner examined after 56 months of exposure was in good condition; properties had changed very little since the first specimen was examined at one year of exposure. An optimal compacted thickness reported for a pond holding primarily water was two layers of 5 cm each, for a total thickness of 10 cm (Matrecon 1980). The quality of the finished liner depends on the compaction during placement. The liner should be compacted to at least 97% of the density, or less than 4% voids. Before placement of the liner, the subgrade should be properly prepared. The subgrade should not have side slopes greater than 2:1 and preferably no greater than 3:1. The soil should be treated with a suitable soil sterilant to prevent puncture of the liner by weeds and roots.

Asphalt liners are effective water resistant material. They are however, not resistant to organic solvents and chemicals, particularly hydrocarbons in which they are partially or wholly soluble. Consequently, asphalts are not effective liners for disposal sites containing petroleum derived wastes or petroleum based compounds such as oils, fats, or aromatic solvents. Asphalt does show good resistance to inorganic chemicals and low permeability to corrosive gases such as hydrogen sulfide and sulfur dioxide. Aggregate containing carbonates should be avoided if wastes are acidic.

Prefabricated asphalt panels have also been used as liners for water reservoirs. These panels typically consist of a core layer of blown asphalt blended with mineral fillers and reinforcing fibers. The panels are 3 to 13 mm thick, one to 1.5 m wide, and 3 to 6 m long (The Asphalt Institute 1976). When properly installed and seamed, the panels form a relatively smooth watertight seal exhibiting the resistance characteristic of asphalt concrete.

8.2.2.2 Soil Cement

Soil cement is a mixture of portland cement, water, and selected soils. The mixture is a low strength concrete with greater stability than natural soil. A fine grained soil produces soil cement with a permeability of about 10^{-6} cm/s. To reduce the permeability of soil cement, coatings of epoxy asphalt and epoxy coal-tar have been used.

Any soil (except organic soils) with less than 50% silt and clay is suitable for soil cement. A high clay content impairs the ability to form a homogeneous material, thus, reducing the efficiency of producing an impermeable layer. The important factors in developing soil-cement liners are cement content, moisture content, and degree of compaction. Soil cements can resist moderate amounts of alkali, organic matter and inorganic salts. The aging and weathering properties of soil cements are good, especially those with wet-dry, and freeze-thaw cycles. As a landfill liner, the main problem of cracking when dry will not arise.

8.2.2.3 Soil Asphalt

Soil asphalt is a mixture of soil and liquid asphalt. A silty gravelly soil with 10–25% silty fines is the preferred soil type. The permeability of soil asphalt after compaction varies with the percent compaction and the percent asphalt. At a high void content (3–10%), soil asphalt has a measurable permeability. Soil asphalts with lower asphalt content are not recommended as lining material. Soil asphalt made with asphalt emulsion is not sufficiently impermeable, and requires a weatherproof hydrocarbon resistant bituminous seal (The Asphalt Institute 1976).

8.2.3 SPRAYED-ON LININGS

Liners for waste disposal sites can potentially be formed in the field by spraying onto a prepared surface a liquid which then solidifies to form a continuous membrane. Such liners have been used in canals, small reservoirs, and ponds. Most of the experience with this type of liner has been with air-blown asphalt. However, a variety of new materials such as bituminous seal coat, emulsified asphalt, urethane modified asphalt, rubber and plastic in liquid and latex form, and organic polymers are becoming available for construction of sprayed-on liners (Loehr 1987). These liners are seam-free, but preparing them pin hole-free in the field poses some difficulties.

The sprayed-on linings are classified into four types: air blown asphalt, emulsified asphalt, urethane modified asphalt, and rubber and plastics in either liquid or latex form.

8.2.3.1 Air-blown Asphalt

Membranes of catalytically-blown asphalt are the most commonly used sprayed-on linings. The asphalts used in making these membrane linings have high softening points and are manufactured by blowing air through the molten asphalt at temperatures in excess of 260°C (500°F) in the presence of a catalyst such as phosphorus pentoxide or ferric chloride. To prepare the membrane, the asphalt is sprayed on a prepared soil surface at a temperature of 200°C (400°F), at a pressure of 340 kP (50 psi) through a slot-type nozzle, and at a rate of about 7 L/m² (1.5 gal yd²). The membrane is formed by making two or more passes of the spray device and overlapping sections by 0.3 to 0.6 m (1–2 feet). Sprayed-on membranes retain their tough flexible qualities indefinitely when properly covered and protected from mechanical damage. The actual placing of the earth covers on a sprayed-on membrane may cause some damage to its integrity. Studies have shown that the addition of 3–5% rubber improves the properties of the asphalt by inducing greater resistance to flow, increased elasticity and toughness, decreased brittleness at low temperatures, and greater resistance to aging (Chan et al., 1978). Except for resistance to hydrocarbon solvents, oils and fats, the chemical resistance of asphaltic liners is quite good.

8.2.3.2 Emulsified Asphalt

Emulsions of asphalt in water can be sprayed at ambient temperatures (above freezing), to form a continuous membrane of asphalt after breaking of the emulsion and evaporation of the water. The membranes are less tough and have lower softening points than membranes of hot applied catalytically-blown asphalt. Toughness and dimensional stability can be achieved by spraying asphalt emulsions onto a supporting fabric. Fabrics of woven jute, woven or nonwoven glass fiber, and nonwoven synthetic fibers have been used with various anionic or cationic asphalt emulsions to form linings.

8.2.3.3 Urethane Modified Asphalt

A urethane modified asphalt liner system is applied onto a prepared surface. A premix is combined with an activator, and sprayed on at a rate of approximately 1.3 L per minute, and covering about one m² (0.28 gal/min.

yd^2). The final membrane is generally recommended to have a thickness of 1.3 mm (50 mil), usually obtained by applying one coat of 1.3 L per m^2 (0.28 gal/yd^2) on horizontal surfaces, or two coats on vertical surfaces. The second coat may be applied about 15 minutes after the first coat. The membrane must be cured for 24 hours before being put into service. The liner material should be applied only to properly prepared surfaces. The surface must be clean and dry. Porous surfaces should be filled. Generally, a primer and a bonding agent is applied prior to the application of the actual membrane. The procedures for preparing base surface, and the necessary precautions are provided by the manufacturer.

8.2.3.4 Rubber and Plastic Latexes

Rubber and resin latexes have also been used for making in situ liners. Polymers of rubber and plastic latexes are sprayed on, allowed to soak into the soil and then permitted to dry. They form a film on the soil surface and produce a good seal.

8.2.4 SOIL SEALANTS

The permeability of some soils, soil cement, and other prepared surfaces can be significantly reduced by the application of various chemicals or latexes. These chemicals may be waterborne, mixed in place, spray applied, or injected below the soil surface (Bureau of Reclamation 1963). Waterborne or spray-on polymer soil sealants can reduce permeability of earth lined impoundments. However, the sealing effect is confined to the upper few centimeters and can be significantly diminished by the effects of wet-dry and/or freeze-thaw cycles. Types of sealants include resinous polymer-diesel fuel mixtures, petroleum based emulsions, powdered polymers which form gels, and monovalent cationic salts (Matrecon, Inc. 1980). Polymeric soil sealants may be applied as a dry blend which is mixed into the soil and compacted, sprayed on as a slurry, or dusted on as a powder. The limitations of the polymer seals are: (1) the polymer itself does not supply strength, therefore, the site must be compacted, and (2) exposure to salts, acids, and multivalent cations causes the polymers to shrink, increasing the permeability and decreasing the effectiveness of the seal.

8.2.5 POLYMERIC FLEXIBLE MEMBRANE LINERS (FMLS)

Polyvinyl Chloride (PVC) liners for swimming pool application were introduced in the early thirties. The geomembrane era began in the 1950s with reservoir liners made from butyl rubber. These thermoset elastomers had been developed during World War II in the search for synthetic rubber. With a growing number of formulations and applications, several names have been given to these liners. Among these names are pond liners, flexible membrane liners (FMLs, the term used by EPA), synthetic membrane liners (SMLs), geomembranes and others (Koerner et al. 1991). Today, the most frequently used liner for waste containment is high density polyethylene (HDPE), primarily because of its high chemical resistance to wastes and leachates.

Prefabricated liners consisting of polymeric sheet (polymeric membranes) can be used for landfills and waste storage impoundments. Such liners can have very low permeabilities and have been used in water reservoirs, sanitary landfills, and other waste facilities. A wide variety of liner materials are being manufactured and marketed. These materials have different physical and chemical characteristics, methods of installation, interaction with specific waste constituents, and costs. Some liners have fabric reinforcement. Detailed information on the applicability of a polymeric membrane should be obtained from the manufacturer. An excellent source of information on the development and implementation of geosynthetic materials is provided by Koerner (1984). General discussion partly adapted from Matrecon, Inc. (1980, 1988) is provided below.

8.2.5.1 Classification of Polymers

Polymers are macromolecular structures formed (polymerized) by chemical union of many units of a specific chemical configuration. The result is a long molecular structure of this repeat unit. The polymers are either inorganic or organic, with the later being more prevalent. The organic polymers are classified as natural, semisynthetic, and synthetic. Bright (1992) provided a generalized classification of polymers with examples of the more commonly known polymers within each class (Table 8.3.).

8.2.5.2 Polymers Used for Manufacture of FMLs

Many types of polymers are used for making FMLs. Each polymer produces a FML that is suitable for a specific application. The physical and chemical properties of polymeric FMLs vary considerably, as do methods of installation and seaming, costs, and interaction with different wastes. The polymers are always used with a variety of ingredients such as fillers, plasticizers or oils, antidegradants, and curatives to improve the selected

TABLE 8.3. Classification of Polymers.

Polymer Class	Examples
Inorganic	Siloxanes
	Silicones
	Sulfur Chains
	Black Phosphorus
Organic	
Natural	Polysaccharides
	Starches
	Cellulose
	Proteins
	Insulin
	Deoxyribonucleic Acid (DNA)
Semisynthetic	Rayon
	Methylcellulose
	Cellulose Acetate
	Starch Acetate
Synthetic	
Thermoset	Epoxies
	Phenolics
	Rubber and Elastomers
	Unsaturated Polyesters
Thermoplastic	Polyolefins
	Vinyl Polymers
	Saturated Polyesters
	Engineering Polymers
	Fluorocarbons
	Alloys and Blends

Source: Bright (1992).

properties depending on end-use, and to reduce the costs. In vulcanization, sulfur crosslinks are formed between the molecules of rubber. The material thus becomes insoluble in solvents, and less susceptible to change in properties with change in temperature. The properties of a polymeric FML also depends on its thickness, fabric reinforcement and type, and number of plies. Most polymeric FMLs are now based on uncrosslinked compounds and, therefore, are thermoplastic. Thermoplastic films are preferred because they are easier to seam and repair effectively during installation in the field. Thermoplastic FMLs can be seamed by various heat sealing or welding methods. If they are noncrystalline, they can be seamed with various adhesives and solvents. A solvent containing dissolved liner compound increases the viscosity and reduce its rate of evaporation during sealing. Common polymers used in manufacture of FMLs are presented in Table 8.4. The composition, general properties and characteristics of many of the polymers is provided below.

TABLE 8.4. Polymers Used in Manufacture of FMLs.

Polymer	Type of Compound Used in FMLs		Fabric Reinforcement	
	Thermoplastic	Crosslinked	With	Without
Butyl rubber	No	Yes	Yes	Yes
Chlorinated polyethylene (CPE)	Yes	Yes	Yes	Yes
Chlorosulfonated polyethylene (CSPE)	Yes	Yes	Yes	Yes
Elasticized polyolefin (ELPO)	Yes	No	No	Yes
Elasticized polyvinyl chloride (PVC-E)	Yes	No	Yes	No
Epichlorohydrin rubber (CO, ECO)	Yes	Yes	Yes	Yes
Ethylene propylene rubber (EPDM)	Yes	Yes	Yes	Yes
Neoprene (chloroprene rubber-CR)	No	Yes	Yes	Yes
Polyethylene (PE)	Yes	No	No	Yes
Polyvinyl chloride (PVC)	Yes	No	Yes	Yes
Thermoplastic elastomer (TPE)	Yes	Yes	Yes	Yes

Source: Matrecon, Inc. (1988).

8.2.5.2.1 Butyl Rubber

Liners of butyl rubber were among the first synthetic liners to be used for potable water impoundments and have been in this type of service for about 30 years. Butyl rubber is a copolymer of isobutylene (97%) with small amounts of isoprene introduced to furnish sites for vulcanization. Vulcanization is a process of heating crude rubber with sulfur or its compounds, and subjecting it to heat in order to make it nonplastic, and increase its strength and elasticity. The important properties of butyl rubber for its use as a liner are: (a) low gas and water vapor permeability, (b) thermal stability, (c) ozone and weathering resistance, (d) chemical and moisture resistance, and (e) resistance to oils and fats. Butyl rubber is generally compounded with fillers and some oils and vulcanized with sulfur. Butyl rubber has a high resistance to mineral acids, has a high tolerance for extremes in temperature and retains its flexibility throughout its service life. It has good tensile strength and desirable elongation qualities. Butyl rubber liners are manufactured in both reinforced and in unreinforced versions of 0.5–3 mm thickness. They are difficult to seam and to repair.

8.2.5.2.2 Chlorinated Polyethylene (CPE)

Chlorinated polyethylene (CPE) is produced by a chemical reaction between chlorine and high-density polyethylene. Presently available polymers contain 25–45% chlorine. Since CPE is a completely saturated polymer it is not susceptible to ozone attack and weathers well. The polymer also has good tensile and elongation strength. Chlorinated polyethylene is characterized by resistance to deterioration by many corrosive and toxic chemicals. Because they contain little or no plasticizer, CPE liners have good resistance to growth of mold, mildew, fungus, and bacteria.

8.2.5.2.3 Chlorosulfonated Polyethylene (CSPE)

Chlorosulfonated polyethylene is a family of polymers prepared by reacting polyethylene in solution with chlorine and with sulfur dioxide. Presently available polymers contain from 25–43% chlorine and from 1.0–1.4% sulfur. Chlorosulfonated polyethylene (CSPE) is characterized by ozone resistance, light stability, heat resistance, good weatherability, and resistance to deterioration by corrosive chemicals, such as acids and alkalies. It is resistant to growth of mold, mildew, fungus, and bacteria. Membranes of this material are available in both vulcanized and thermoplastic forms. Usually, they are reinforced with a polyester or nylon scrim. The fabric reinforcement provides needed tear strength to the sheeting for use on slopes, and reduces the distortion resulting from shrinkage when placed on the base and when exposed to the heat of the sun. Membranes of this polymer do not crack at extremes of temperatures or from weathering. Disadvantages of CSPE membranes are low tensile strength, and has a tendency to shrink from exposure to sunlight. It has relatively poor resistance to oils, and some CSPE tends to harden on aging.

8.2.5.2.4 Elasticized Polyolefin (ELPO)

Elasticized polyolefin is a blend of rubbery and crystalline polyolefins. This polymeric material, introduced in 1975 as a black unvulcanized, thermoplastic liner, is readily and easily heat sealed in the field or at the factory using a specially designed heat welder. It has a low density (0.92) and is highly resistant to weathering, alkalis, and acids. The sheets of membrane (6 m wide and 60 m long) are shipped to the site for assembly in the field.

8.2.5.2.5 Epichlorohydrin Rubbers (CO and ECO)

This classification includes two epichlorohydrin-based elastomers which are saturated, high molecular weight, aliphatic polyethers with chloromethyl side chains. These materials are vulcanized with a variety of reagents. Epichlorohydrin elastomer vulcanizates exhibit the following characteristics: (a) resistance to hydrocarbon solvents, fuels and oils, (b) ozone and weathering resistance, (c) low rate of gas/vapor permeability,

(d) thermal stability, and (e) good tensile and tear strength. Epichlorohydrin rubber has a high tolerance for temperature extremes and retains its flexibility at extreme temperatures throughout its service life. Epichlorohydrin elastomers can be seamed at room temperature with vulcanizing adhesives.

8.2.5.2.6 Ethylene Propylene Diene Mono Rubber (EPDM)

Ethylene propylene rubber is a copolymer ethylene, propylene and a minor amount of nonconjugated diene hydrocarbon. These rubbers vary in ethylene and propylene ratio. Liners of EPDM compounds have excellent resistance to weather and ultraviolet exposure, and when compounded properly, resist abrasion and tear. Because of its excellent ozone resistance, small amounts of EPDM are sometimes added to butyl rubber to improve its weather resistance. EPDM liners tolerate extremes of temperature, and maintain their flexibility at low temperatures. They are resistant to dilute concentrations of acids, alkalis, silicates, phosphates, and brine, but are not recommended for petroleum solvents (hydrocarbons), or for aromatic or halogenated solvents. EPDM liners are supplied in vulcanized sheeting of 0.5–3 mm thickness, both supported and unsupported. These sheets are seamed by special cement and require careful application to assume satisfactory field seaming.

8.2.5.2.7 Neoprene

Neoprene is the generic name of synthetic rubbers based upon chloroprene. These rubbers are vulcanizable, usually with metal oxides, but also with sulfur. They closely parallel natural rubber in mechanical properties, e.g., flexibility and strength. However, neoprene is superior to natural rubber in its resistance to oils, weathering, ozone, and ultraviolet radiation. Neoprene is resistant to puncture, abrasion, and mechanical damage. Neoprene membranes have been used primarily for the containment of wastewater and other liquids containing traces of hydrocarbons. Neoprene sheeting for liners is vulcanized, therefore, vulcanizing cements and adhesives must be used for seaming.

8.2.5.2.8 Polyethylene

Polyethylene is a thermoplastic polymer based upon ethylene. Polyethylene resins have long been described on the basis of their density as low-density (LDPE), medium-density (MDPE), and high-density (HDPE) (ASTM 1992). The properties of polyethylene liners largely depend on their crystallinity and density. The HDPE liners commonly used in a wide range of waste containment applications are produced from medium density polyethylene resins with density of 0.935–0.940 g/cc. The latest form of polyethylene resin developed for use in geomembrane industry is very low-density polyethylene (VLDPE) with density of 0.900–0.910 g/cc. The VLDPE liners are specifically used for landfill caps because of their higher flexibiltiy and elongation characteristics and excellent cold temperature resistance.

Polyethylene liners are inert and consist of 97% polyethylene and 3% carbon black and chemical stabilizers, and do not contain any plasticizers or fillers. Polyethylene liners are ultraviolet light resistant and are suitable material for exposed and buried applications. Polyethylene liners are not affected by microorganism, fungi, or rodent attack. The primary seaming method for polyethylene liners is hot shoe fusion welding. Extrusion welding is used for repair work. Polyethylene liners are the most common synthetic liners used in waste containment applications.

8.2.5.2.9 Polyvinyl Chloride (PVC)

Polyvinyl chloride is produced by any of several polymerization processes from vinyl chloride monomer. It is a thermoplastic polymer which is compounded with plasticizers and other modifiers to produce a wide range of physical properties. Elasticized polyvinyl chloride (PVC-E) has elasticizer such as dioctalthallate adsorbed into it. The resulting product has improved flexibility, elongation, and tear strength characteristics.

PVC liners are produced in roll form in various widths and thicknesses. Most liners are used as unsupported sheeting, but fabric reinforcement can be incorporated. PVC compounds contain 25% to 35% of one or more plasticizers to make the sheeting flexible and rubber-like. They also contain 1% to 5% of a chemical stabilizer and various amounts of other additives. Plasticizer loss during service is a source of PVC degradation. The loss of plasticizer may occur due to volatilization, extraction, and microbiological attack.

Plasticized PVC sheeting has good tensile, elongation, and puncture and abrasion resistance properties. It is readily seamed by solvent welding, adhesives, and heat and dielectric methods. PVC membranes are the most widely used of all polymeric membranes for waste impoundments. They show good resistance to many inorganic chemicals.

8.2.5.2.10 Thermoplastic Elastomers (TPE)

Thermoplastic elastomers are manufactured by blending natural or synthetic elastomers such as rubber, block polymers or polyolefins with thermoplastic materials (e.g., polypropylene). Thermoplastic products have great flexibility in the way they may be manufactured, and a variety of properties may be obtained by varying the blending proportions or by using additives. Thermoplastic sheets may be seamed either using heat or with solvents. TPEs can repeatedly be stretched to twice their relaxed length. Therefore, they exhibit great resistance to tearing or puncture.

8.2.5.3 Manufacturing

FMLs are manufactured by three different processes. These are calendaring, extrusion, and spread or knife coating. Calendaring is used in forming both unreinforced and fabric-reinforced sheeting, whereas extrusion is only used in making unreinforced sheeting. Spread coating is used for making fabric-reinforced sheeting. Matrecon, Inc. (1988) has provided detailed discussion on FMLs manufacturing, and methods of vulcanization, seaming, welding, splicing and repairing.

8.2.5.4 Commonly Used Geomembranes (FMLs)

Koerner et al. (1991) reported that geomembranes currently in common use in waste impoundment can be grouped into three broad categories: semicrystalline, flexible, and reinforced. All three types may be seamed using thermal methods. These types are briefly discussed below (Koerner et al. 1991).

8.2.5.4.1 Semicrystalline

Semicrystalline geomembranes are made from thermoplastic materials, and are characterized by stiffness values in excess of 1,000 gm/cm, which distinguish them from the flexible and reinforced geomembranes. The most commonly used semicrystalline liner materials are smooth, and coextruded or textured sheets of high density polyethylene (HDPE). HDPE is manufactured from polyethylene resin, with carbon black (about 2 percent) added for protection from ultraviolet radiation. Trace amounts of antioxidants and other proprietary additives and processing agents are also added.

8.2.5.4.2 Flexible

Flexible thermoplastic geomembranes, sometimes called low crystalline thermoplastics, are defined by stiffness values below 1,000 gm/cm. A number of different materials are used to construct flexible geomembranes, which are in turn manufactured from a wide variety of resins, polymers, plasticizers, fillers, carbon black and additives. Common materials used include very low density polyethylene (VLDPE), polyvinyl chloride (PVC), chlorinated polyethylene (CPE), and chlorosulfonated polyethylene (CSPE). VLDPE liners do not contain any plasticizers or fillers.

8.2.5.4.3 Reinforced

Reinforced geomembranes combine flexible thermoplastic materials with needle punched or woven fabrics such as nylon, polyester, polypropylene or glass fiber inbedded in the membrane during calendaring, or spread-coated. The resulting product exhibits greater strength and thermal stability than does its unreinforced counterpart. Most commonly used reinforced thermoplastic geomembranes are reinforced chlorinated polyethylene (CPE-R), reinforced chlorosulfonated polyethylene (CSPE-R), and reinforced ethylene interpolymer alloy (EIA-R).

Other types of geomembranes are also being developed; a wider variety will no doubt be available in the future.

8.2.6 COMPOSITE LINERS

A composite liner is composed of an engineered soil layer overlain by a synthetic flexible membrane liner. This combination provides higher protection than individual liners because each liner component has different resistance properties. Discussion on composite liners may be found in Sec. 3.8.1.

8.3 LINER SELECTION AND PERFORMANCE

Natural and synthetic liners may be utilized as both a collection device, and as a means for isolating leachate within the fill to protect the soil and groundwater below. Of concern is their ability to maintain integrity and impermeability over the life of the landfill. Subsurface water monitoring, leachate collection, and/or clay liners commonly are included in the design and construction of a waste landfill when polymeric membranes are used.

8.3.1 LINER EVALUATION

To effectively serve the purpose of containing leachate in a landfill, a liner system must possess a number of physical properties. Kumar and Jedlicka (1973) have provided many criteria for synthetic liner selection. Some of these criteria are listed on the following page:

- high tensile strength, flexibility, elongation without failure
- resists abrasion, puncture, chemical degradation by leachate
- good weatherability, manufacturer's guarantee for long life
- immune to bacterial and fungal attack
- color: black (to resist UV light)
- minimum thickness, 20 mils
- membrane should have uniform composition, free of physical defects
- withstand temperature variation and ambient conditions
- easily installed
- economical

The natural soil containment system often provides higher permeability, but, with sufficient depth, may have the capacity to attenuate the contaminants. Synthetic liners have been under intensive evaluation to establish their structural strengths, chemical reactions with different wastes and physical properties. Volumes of literature exist on this subject. Laboratory and field experiments have been conducted with mixtures of wastes and liners to develop and estimate the compatibility of various liners (Kumar and Jedlicka 1973; Ewald 1973; Haxo 1976, 1977, 1978; Haxo et al. 1985; Haxo and Caney 1988; Matrecon, Inc. 1980, 1988; Koerner et al. 1991). Comparative rating of natural soil and admix materials used for various industrial waste impoundments are provided in Table 8.5. Chemical resistance information on selected polymers used in FMLs is provided in Figure 8.2 (Poly-Flex Inc. undated). The physical, and structural properties of FMLs and admixed after exposure to landfill leachate are provided in Table 8.6 and Table 8.7, respectively. Ideally, the final selection of any liner material should be based on exposure tests. Manufacturers will provide basic properties, and information regarding approximate resistance to chemical attack.

The estimated service life of a liner in a particular exposure condition is an important factor in selecting a liner material. For temporary holding situations such as impoundments and waste piles, a short life may be satisfactory. For a hazardous waste landfill a very long service life is required. Principal causes of liner failure are:

(1) *Physical*—puncture, tear, differential setting, thermal stress, hydrostatic stresses, abrasion, cracking

(2) *Chemical*—solvents, hydrolysis, acids, bases, chemical oxidation

(3) *Biological*—microbial degradation

Physical failures of FMLs are commonly related to poor subgrade surface preparation, resulting in rocks puncturing the liner. Other failures are due to faulty seams and excessive stresses on the liner system. Chemical

TABLE 8.5. Liner-Waste Compatibilities.

Liner	Caustic Petroleum Sludge	Acidic Steel-making Waste	Electroplating Sludge	Toxic Pesticide Formulations	Oil Refinery Sludge	Toxic Pharmaceutical Waste	Rubber & Plastic Waste
Soils							
Compacted clay	P	P	P	G	G	G	G
Admixes							
Asphalt							
concrete	F	F	F	F	P	F	G
Asphalt							
membrane	F	F	F	F	P	F	G
Soil asphalt	F	P	P	F	P	F	G
Soil cement	F	P	P	G	G	G	G
Soil bentonite	P	P	P	G	G	G	G

P = poor, combination should be avoided; F = fair, combination should be tested before use; G = good, probably satisfactory.
Source: U.S. EPA (1983)

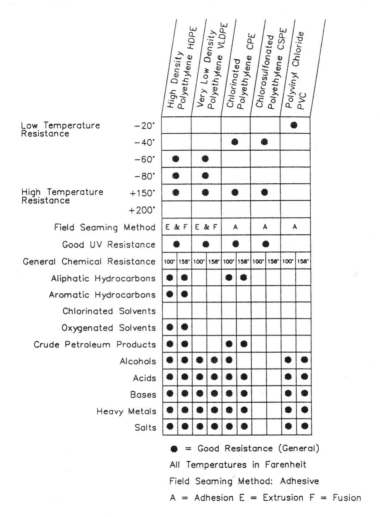

Figure 8.2 Chemical resistance of polymers used in common FMLs. Courtesy Poly-Flex, Inc.

failures normally are related to the characteristics of the waste in contact with the liner. Microbial degradation occurs due to growth of bacteria or fungi that may solubilize the fillers and plasticizers.

Synthetic liners are thin sheets (0.5 to 2.0 mm thick) composed of materials such as rubber, polyvinyl chloride, or various polyethylenes. Most synthetic liners are considered impermeable to water, especially when compared with natural soil liners. Some laboratory experiments suggest that volatile organic compounds (e.g., trichloroethylene, toluene, and xylene) can migrate rapidly through synthetic liners (Haxo and Caney 1988). The amounts and directions of migration vary depending on the

TABLE 8.6. Comparison of Physical and Structural Properties of 30 Mil HDPE, VLDPE, and PVC.

Property	Test method	HDPE	VLDPE	PVC
Gauge, mils	ASTM D 1593	30	30	30
Density (gm/cc)	ASTM D 1505	0.950	0.925	1.2–1.4
Tensile Strength@ Yield (lb/in)	ASTM D 638	75	*	*
Elongation@Yield (%)	ASTM D 638	10	*	*
Tensile Strength@ Break (lb/in)	ASTM D 638	125	110	90
Elongation@ Break (%)	ASTM D 638	800	1000	300
Secant Modulus 1% Elong. (psi)	ASTM D 638	90,000	15,000	5–10,000
Tear Strength (lb)	ASTM 1004	22	14	8
Puncture Resistance (lb)	FTMS 101 C 2065	50	45	42
Dimensional Stability, %	ASTM D 1204 212°F, 15 min	±1	±3	±10
Moisture Vapor Transmission Rate g/m²·day	ASTM E 96 100°F @ 100% relative humidity	0.43	0.77	7.30
Cold Temperature Brittleness	ASTM D 746	< −94°F	< −94°F	< −20°F

*VLDPE and PVC do not exhibit true yield points.
1 mil = 10^{-3} in or 0.025 mm; 1 lb = 453.6 g; 1 lb/in² = 6.89 kN/m²; 1°F = (1.8 × °C + 32).
Source: Courtesy Poly-Flex, Inc.

TABLE 8.7. Properties of Admixed and Asphalt Liners.

	Asphalt Concrete	Soil Cement	Soil Asphalt
Properties of unexposed samples			
Density, g/cm³	2.39	2.17[a]	2.23
Compressive strength			
Initial, psi	2805	1910[a]	1220
After 24 h immersion in water[b], psi	2230	320[a]	185
Retention of strength after immersion, percent	80	69[a]	15
Properties after 56 months of exposure[c]			
Density, as received, g/cm³	2.54	—	2.31
Density, dry, g/cm³	2.417	—	2.02
Compressive strength, psi	258	1182	26
Percent of original	9	62	2

[a]Measured on specimen molded according to ASTM D558 (1992). Wet density, 2.35 g/cm³ (146.4 lb/ft³).
[b]Asphalt concrete immersed in water at 60°C; soil cement and soil asphalt immersed at room temperature.
[c]Measured on 2-inch (5-cm) cores cut from liner specimens.
Source: Adapted in part from Haxo et al. (1985), and Matrecon, Inc. (1980, 1988).

characteristics of the chemicals, the liner material and thickness, temperature, and concentration of the chemicals on either side of the liner (Matrecon, Inc. 1988). In different tests conducted by EPA, most synthetic liners were eventually destroyed when exposed to methylene chloride in full strength concentrations (Curran and Frobel 1985). It is not known whether these results are representative of actual landfill conditions. In laboratory experiments reported by Haxo and Caney (1988), the concentrations of organic chemicals in the solutions were relatively high; for example, trichloroethane (TCE) was 1,100 ppm. In a landfill, synthetic liners probably encounter more dilute solutions in most cases. The average concentration of TCE in leachate of several municipal landfills was generally less than 200 ppm (Table 6.5.). Thus, laboratory experiments with immersed liners involve high concentrations of chemicals that may not represent field conditions.

An important aspect of flexible membrane liners is the process by which the seams of the different liner segments are joined. Segments of a liner can be joined together in the factory by using solvent adhesives or dielectric methods, or in the field using various welding methods. Morrison and Parkhill (1987) have tested seam strength under conditions designed to simulate chemical and physical environments that might be encountered at hazardous waste facilities. Two types of strength generally were evaluated: peel strength (i.e., the ability of the seam to resist peeling apart of two liner segments), and shear strength (i.e., the ability of the liner material to resist lateral separation). Results indicated that shear and peel strength are not correlated. Peel strength is related to the strength of the bond, and shear strength is related to the properties of the liner material. The method used to create the seams causes differences in peel and shear strengths. Morrison and Parkhill (1987) concluded that existing data and manufacturers' recommendations on the chemical compatibility of the liner materials provide an initial basis for evaluating expected liner performance in given chemical environments. However, tests of less than 6 months may be inadequate to determine the performance of some flexible membrane liners.

8.3.2 IMPROVED LINER DESIGN

Subtitle D regulations require liners for landfill construction (Chapter 4). Woods (1992) reported that in anticipation of a large potential market for liners/geotextiles especially in municipal landfills, the liner manufacturers have developed, and are marketing improved products. Improvements include less cracking, more strength and durability, more conformity to irregularities found at the base of the landfill, and stronger seams. Manufacturers are producing sheets of high density and very low density polyethylene that have greater crack resistance and stability.

Many liners are wider (over 10 m wide) which means fewer seams. Liners with a high coefficient of friction allow placing on steeper slopes. Under Subtitle D regulations, liner installations require 0.6 m soil, and an additional layer of granular material that must accompany the flexible membrane liner. With thick and high strength synthetic geotextiles, these layers can be safely reduced thus allowing more space for solid wastes. Woods (1992) reported one such liner that is almost 2 cm (0.75 inch) thick and can withstand pressures of 550 kPa/m² (80 psi), or an equivalent of 50 m (164 ft) head of water. Other laminated composite materials provide better filtration, and leachate flow characteristics that facilitate design for collection systems (Woods 1992).

8.4 LINER CONSTRUCTION AND INSTALLATION

The planning, design and construction of a lined sanitary landfill is an engineering operation requiring quality control and workmanship. The construction methods are different for different liner categories. These categories are presented in Sec. 8.2. In all cases the following steps are needed: (1) subgrade preparation, (2) liner installation, and (3) special facilities.

The subgrade preparation involves site work such as excavation and grading, subsoil compaction, and sterilization. Soil sterilization by a selected herbicide is sometimes necessary to control the growth of plants that may damage the liner. Liner installation requires equipment and material storage, installation, inspection and quality assurance program. Special facilities involve installation of under drainage systems, leachate collection and removal systems, and monitoring wells.

8.4.1 SOIL AND CLAY LINER

The main objective of soil and clay liners is to impede the downward movement of the liquid into the underlying, undisturbed soil. The desired soil composition, depth, and densification must be achieved and compaction procedure must be followed to achieve permeability less than the recommended value. The principal requirements of the soil liner used are permeability, adsorptive capacity, and compatibility with waste and structural strength. Determination of atterberg limits, permeability, compatibility, and chemical sensitivity must be made to develop procedure for liner preparation. Bentonite clay or other material if necessary may be mixed in proper proportion with the soil to achieve the desired properties of the soil liner.

8.4.2 ADMIXED LINER

Admixed liners are formed from soil cement, soil concrete, cement concrete, asphalt concrete, and asphalt panels. In all cases, the basic requirement is subgrade preparation, placing of the admixed material, and construction of the liner as specified by the designer. The subgrade should be smoothed by rollers after compaction. If water table is high, a layer of gravel or drainage system should be placed beneath the liner. The grade should be well moistened before placing the soil cement or concrete. A prime coat of hot asphalt should be applied to the surface for asphalt concrete liner. The soil cement, or cement concrete is placed using road paving methods and equipment. The compaction and curing procedures must be properly followed. The hot asphalt concrete mix should be placed, spread and compacted to achieve smooth surface free from grooves, depressions, and holes. Joints should be made as per specifications. The prefabricated asphalt panels require careful planning and workmanship for proper installation and jointing.

8.4.3 BOTTOM SEALERS

Bottom seal is a method by which sealing material is applied on the bottom of a new landfill site, or a pressure-injected-grout is pumped under the existing landfill site. A variety of materials, such as asphalt concrete, soil asphalt, soil cement, sprayed asphalt membranes, and bituminous seals may be utilized for sealing purposes depending upon the soil properties (Lu et al. 1985). The bottom seal prevents groundwater from entering the landfill, and also impounds the leachate collected within the fill. The disadvantage of pressure-injected grout under the existing landfill is uncertainty of a complete seal.

8.4.4 SYNTHETIC MEMBRANE LINERS

Synthetic membrane liners do not require as much intensive labor as clay or admixed liners. Nevertheless, extreme care is necessary for installation. The subbase must be carefully prepared. Stone gravel and hard pieces larger than 1.25 cm (0.5 in) must be removed. Soil sterilization may be necessary. The liner installation may require the following care:

(1) The base must be properly contoured and sloped toward the leachate collection points. A slope of 2% or more is desirable. The construction details of leachate collection and removal system is provided later in Sec. 8.5.

(2) Prior to liner installation, the exposed subgrade should be scored to a depth of at least 15 cm (6 inches) and compacted to 95% of standard proctor (ASTM D 698), and to the maximum laboratory dry density. Weak or compressible areas which cannot be satisfactorily compacted should be removed and replaced with properly compacted fill. All surfaces to be lined should be smooth, free of all foreign and organic material, sharp objects, or debris of any kind. The surfaces should provide a firm, unyielding foundation with no sharp changes or abrupt breaks in grade. Standing water or excessible moisture should not be allowed.

(3) Laying of the liner should be avoided during winds higher than 24 km/h (15 mph).

(4) The temperature range for liner installation should be 4–27°C (40–80°F). This is necessary because the liners absorb heat and become overheated. This may result in wrinkles due to seaming at different times of the day, especially if temperature variations are large.

(5) No vehicle or equipment except seaming equipment, seam testing equipment, and only a minimum number of personnel should be allowed to move on the liner. Bagchi (1990) suggested that a small piece of synthetic membrane, placed below the membranes that are being seamed, may reduce burnout due to small depressions. This piece is moved forward along with the seaming equipment. The crew should be allowed to carry only necessary tools, and should not wear heavy boots.

(6) Welding or seaming should not be done if it is raining, or the membrane is wet. In the mornings, this operation should be started only after the dew has evaporated. If seaming must be done in wet weather, proper protective cover should be used to keep the operation dry.

(7) The field seams should be nondestructively tested for leaks, and destructively tested for seam strengths, peel adhesion, and shear strength. The synthetic membranes must be covered with soil or drainage layer as soon as possible. It is necessary that enough material must be stock piled for covering the membrane soon after installation. The synthetic membrane can be damaged by hoofed animals. The thickness of soil covering the membrane should be 0.3 m (12 inch) of sand or similar soil. The soil moving heavy equipment must be carefully used as damage of membrane due to traffic can be severe and may remain undected. The liner installation details are shown in Figure 8.3.

(a)

(b)

Figure 8.3 Liner installation details, (a) prepared base, geonet, and HDPE geomembrane being installed, (b) HDPE geomembrane with anchor. Courtesy Poly-Flex, Inc.

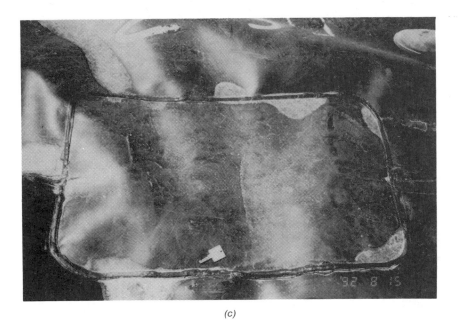

(c)

Figure 8.3 (continued) (c) Typical patch for HDPE extrusion welded to the liner. Courtesy Poly-Flex, Inc.

8.5 LEACHATE COLLECTION AND REMOVAL SYSTEMS

8.5.1 CONSTRUCTION AND OPERATION

Leachate collection and removal systems use pipes to collect the leachate that percolates into the fill. Efficient removal of leachate is essential to prevent ponding, and to relieve hydrostatic pressure over the liner. According to EPA guidelines, the maximum head of leachate over the primary liner should not exceed 30 cm (12 inches). A typical leachate collection system consists of a series of perforated collection pipes (usually PVC pipes of 10 to 15 cm diam), drainage layers and blankets, header pipes, and sumps. The pipes are placed above the liner in drainage layers filled with sand or gravel. In landfills with double liners, the pipes are placed both above the top of the liner, and between the top and bottom liners. In general, liners are designed with a slope so that leachate drains into a central collection point. Landfill base construction, and liner and underdrain details are shown in Figures 3.6.–3.9. Bagchi (1990), and Haxo et al. (1985) also provided many construction details of leachate collection system. The drainage blanked over the liner involves placing a sand layer or a geonet layer to achieve proper drainage. Sand or fine gravel may be used for this purpose. The quality control tests for drainage blanket include grain size analysis and permeability.

An underdrain system below the liner may often be necessary to collect the groundwater for preventing pressure build-up and possible rupture of the liner in the event of a rising water table. Collection piping placed beneath the liners can also facilitate detection and collection of leachate leakage through a ruptured liner, thus minimizing downward percolation of leachate from the landfill.

In an existing landfill which does not have a leachate containment and removal system, construction of such a system often presents a unique engineering problem. One such example was cited by Heglie et al. (1990). The investigators developed the leachate collection system for the Mountain View, California Sanitary Landfill. The system is noteworthy because it involves "retrofitting" a closed landfill with leachate collection systems designed to specific performance objectives, including prevention of groundwater contamination. Most of the refuse cells extended below the groundwater table, and up to 6 m (20 ft) of leachate depth was present in some cells.

Several options were technically evaluated. Among them were: (a) wells provided at desired locations to maintain an inward gradient over the entire cell to remove excess head, (b) perimeter leachate removal using a toe drain or perimeter drain, (c) horizontal borings to install drains, (d) partial or complete slurry walls along cell boundaries, and (e) groundwater extraction from perimeter ditch wells.

Cronin et al. (1990) provided design, operation, maintenance, and cost considerations for leachate collection and removal systems at a sanitary landfill facility in New Jersey. The leachate at this landfill was of high strength (BOD = 22,000 mg/L). Two different leachate collection systems were designed. One system was for an older section which did not have liner requirements. In this section, an interceptor system was constructed of 20 cm (8 in) perforated PVC pipe connected to manholes located a maximum of 110 m (360 ft) apart. The leachate collected by the system was piped to a pumping station. The average leachate production was 133,000 L/d (35,000 gpd). The leachate collection system in the new section of the landfill utilized a clay liner system. A 90-cm (3-ft) thick impermeable clay liner sloped at 1% toward and outer containment dike was provided.

The leachate collection system consisted of 10 cm (4 in) perforated PVC lateral built within an envelope of crushed stone and wrapped with non-woven filter fabric to prevent siltation. The perforations in PVC pipes consisted of 1.25 cm (0.5 in) diameter holes spaced 15 cm (6 in) on centers in three rows along the axis of the pipe separated by 120 degrees. The 10-cm (4 in) laterals ran radially into the center of the landfill and were spaced approximately 76 m (250 ft) apart. The laterals provided an unobstructed flow path for the leachate to the header system. The header system consisted of 20 cm (8 in) perforated PVC lines, also enclosed in stone and filter fabric, and located along the landfill perimeter. The headers conveyed the leachate from the collection laterals to the pump stations. Leachate collection manholes were constructed at the intersection of the 10-cm (4 in) laterals and 20 cm (8 in) perimeter header system and provided access to the system. The future expansions of this landfill utilized a composite liner system. The above design provided structural strength to support maximum anticipated static and dynamic loads and stresses imposed on the pipes. A factor of safety of 1.5 was used. The slope of the collection piping provided a minimum velocity of 0.6 m/s (2 fps).

Maintenance of the piping was necessary to ensure the free flow of leachate. Inspection of leachate lines involved visually observing leachate flow at the manholes and keeping and analyzing records of leachate production. These inspection techniques, however, were not effective for early detection of problems. A noticeable change in leachate flow whether determined visually, or from a review of leachate production logs, were not expected until a significant failure had occurred within the leachate collection system. Remote TV camera inspection was possible for larger lines but was expensive since it usually required the cleaning of the pipeline prior to inspection and access to the pipeline from each end of the pipe.

The leachate lines were cleaned by high pressure pump, hose and jetting nozzle. Pressure as high as 15,200 kN/m² (2,200 psi) was used to break through. The cleaning discharge was black in color indicating possible biological growth in piping system. The material in the piping was predominantly inorganic, having very high concentration of iron and calcium carbonated precipitate.

The cleaning of the leachate system provided useful information which was incorporated into future designs. Future leachate collection systems were designed with a composite liner system, and collection pipes 15 cm (6 in) in diameter were used, as opposed to the 10 cm (4 in) pipes which were installed in the past. This reduced the tendencies toward pipe blockage. A yearly cleaning program for the leachate collection system was instituted.

8.5.2 PERFORMANCE EVALUATION

Many researchers have attempted to analyze the performance of lateral drainage collection systems in sanitary landfills. The steady-state case is of practical as well as of fundamental interest, since approximately steady-state conditions are likely to develop eventually in the leachate collection system. The primary performance criteria are the leakage rate through the liner beneath the drain layer, and the maximum depth of saturation in the drainage layer. McEnroe and Schroeder (1988), gave four primary factors that affect the collection system performance under steady-state conditions:

(1) The rate at which leachate drains from the waste layer into the drain layer (termed the impingement rate)
(2) The distance between the parallel drains
(3) The saturated hydraulic conductivity of the drainage layer
(4) The hydraulic conductivity, slope and thickness of the liner

These investigators analyzed the performance of leachate collection system with low-permeability soil liner under steady-state conditions. Equations and graphs are developed from numerical solution of governing differential equations. Under normal condition, leakage rate is sensitive only to the hydraulic conductivity of the liner.

For a horizontal liner, the shape of the phreatic surface is described by the well-known *Dupuit-Forchheimer* parabola. McBean et al. (1982) presented an approximate solution for steady-state drainage on a sloping impermeable liner based on linearization of the governing equations. Wong (1977) presented a simple approximate solution for unsteady drainage resulting from an instantaneous uniform buildup of head on a sloping low-permeability liner. Kmet et al. (1981), and Demetrocopoulos et al. (1984) used Wong's solution to explore the performance and sensitivity of the collection system to various factors. Lentz (1981) presented another approximate solution for unsteady flow on a sloping low-permeability liner based on linearization of the governing equations. Korfiatis and Demetrocopoulos (1986) solved the complete unsteady flow equations numerically, and compared these results with Wong's approximate solution.

The efficiency of leachate collection systems depends on the rate of leachate generation, spacing of collection pipes, slope of liners, liner permeabilities, and presence of drainage blankets. U.S. EPA (1983) used models to estimate leachate collection efficiencies at 100 L per hectare per day rate (considered typical of landfills). The composite liners exhibited collection efficiencies approaching 100 percent, while systems associated with clay liners would exhibit much lower efficiencies.

Only a small number of existing landfills have any type of leachate collection system, and available data do not allow a determination of how much leachate is actually subject to collection. In addition, the presence of a leachate collection system is not necessarily sufficient to prevent groundwater contamination. U.S. EPA (1988) has identified municipal landfills equipped with leachate collection systems that failed to prevent such contamination because of inadequate design and/or construction.

Once leachate is collected, it can be managed by recirculating it in the landfill, transporting it to a municipal wastewater treatment plant, discharging it to a treatment plant through a sewer, or treating it on site with physical, chemical and/or biological treatment processes. According to U.S. EPA (1988), the discharge of leachate to surface water is expected to decline in the future, while the use of recirculation and transportation to treatment plants is expected to increase.

8.5.3 QUALITY ASSURANCE

Construction quality assurance (CQA), and construction quality control (CQC) are important factors in overall quality management of leachate control system in sanitary landfills and other waste impoundments. CQA involves a planned system of activities to assure that the facility is constructed as specified in the design. CQA includes inspection, verification, audit, and evaluation of materials, products and workmanship. CQC refers to measures taken by the contractor to determine compliance with requirements for materials and workmanship as stated in the plans and specifications for the project.

A written quality assurance plan must be developed and approved by the concerned agency prior to construction. It may be a part of the permit. Complete documentation of activities is essential for those individuals who were unable to observe the entire construction process. Standard daily reporting and documentation, completion of inspection sheets, problem indentification and corrective measures, as built drawings, and final documentation and certification should all be properly stored.

U.S. EPA (1986) has developed a technical guidance document on construction quality assurance for land disposal facilities that provides detailed information on this subject.

8.6 COSTS OF LEACHATE COLLECTION SYSTEM

The construction costs of leachate collection systems depend upon many factors. Among these are liner material, subbase preparation, leachate collection system, type of soil on site, and quality control. In this section the cost of leachate collection systems is presented.

8.6.1 UNIT COSTS OF EMBANKMENT AND MATERIAL

Matrecon, Inc. (1988) provided detailed cost information on the various components of a leachate containment system. Necessary information on these cost estimates was collected from Sai and Zabcik (1985), and E. C.

Jordan Company (1984). Potential cost elements of a waste containment system are listed in Table 8.8. The unit cost of various components may vary greatly depending upon the soil, cost of construction materials, transportation cost, groundwater level, design of containment system, and quality control. The unit cost of major embankment components and drainage pipe are summarized in Table 8.9. Costs have been updated to 1994 values using the *Engineering News Record Construction Cost Index* (ENR-CCI).

8.6.2 INSTALLATION COSTS OF LINERS

The cost of liner components include soil, sand, gravel, and geosynthetic materials. The cost of manufacturing synthetic liners is moderate; their price mainly depends upon the price of raw materials used for manufacture. Other cost items are transportation, storage and installation.

TABLE 8.8. **Potential Cost Elements of Leachate Collection Systems.**

Geotechnical investigation of site
Clearing and grubbing
Excavation volume
Grading and compaction
Berm embankment construction
Compatibility testing of the component materials
Soil component of bottom liner
FML component of composite bottom liner
Components of a secondary leachate collection system:
Drainage layer (synthetic or granular)
Filter layer
Protective soil layer
Geotextile support layer
Leachate collection pipes
FML top liner
Components of a primary leachate collection system
Soil cover above top liner
Auxiliary cleanouts
Pump
Sump
Diversion ditch
Riprap
Quality control and quality assurance

Source: Adopted from Sai and Zabcik (1985).

TABLE 8.9. Unit Costs of Major Embankment Components, Sand and Gravel and Drainage Pipe.

Item	Cost Range, $ 1994	Factors Influencing Costs
A. Major embankment components[a]		
Excavation	1.58–26.35/cy	Project size, soil condition, groundwater, material disposal
Embankment	2.11–21.08/cy	Availability of material, compaction requirements
Drain lines	10.54–52.69/ft	Size, type, backfill requirement
Drain material	8.43–26.35/cy	Size of drain, gradation, availability
Rip rap	15.81–31.62/cy	Availability and distance
B. Media[b]		
Sand (well graded)	$ 6.85/cy	Distance
Grand (well graded)	$ 6.85/cy	Distance
C. Base Construction[b]		
Perforated pipe, 15 cm (6 in)		
PVC (flexible)	1.26–3.43/ft	Material, diameter, distance
HDPE	2.11–8.43/ft	Material, diameter, distance

[a]The original costs were provided by Matrecon, Inc. (1988) in 1987 dollars. These costs have been updated using ENR Construction Cost Indexes (CCI) for 1987 annual average and for March 1993. ENR-CCI for 1987 (average) = 4400.77; ENR-CCI for March 1994 = 5381.04.
[b]The original costs were provided by E. C. Jordan Company (1984) in 1984 dollars. These costs have been updated using ENR Construction Cost Indexes (CCI) for 1984 (annual average) and for March 1994. ENR-CCI for 1984 (average) = 4147.97. ENR-CCI for March 1994 = 5381.04.
   $/cy × 1.31 = $/m^3
   $/ft × 3.28 = $/m

Matrecon, Inc. (1988), provided cost data for selected liner materials based on estimates provided by various manufacturers, fabricators, and installers of the specific materials. Costs are per square foot (m^2) of liner material, installed in quantities sufficient to line a (9300 m^2) 100,000 ft^2 area with a single liner. These costs do not include site and surface preparation, engineering design or soil cover. Cost will be affected by transportation, size of project, time of year, local labor, and complexity of installation. The unit costs also do not represent the leachate collection and detection system. Also, the costs are not equalized to the same service life and performance level. The cost presented by Matrecon, Inc. (1988) have been updated using Engineering News Record Construction Cost Index, and are presented in Table 8.10. These unit costs must be adjusted by adding the cost of subbase preparation, soil cover, leachate collection and detection, and other factors that may apply.

Several sources for cost of landfill leachate collection systems have been developed. A database on leachate collection systems was developed by EPA (1983). Sai and Zabcik (1985) developed a model using *LOTUS* 1-2-3 spreadsheet to calculate design variables and the engineering and construction costs of leachate collection and removal systems.

8.6.3 COSTS OF ADMIXED LINERS

The cost estimates for admix and sprayed-on asphalt membrane liners are provided in Table 8.11. These costs were originally developed by Matrecon, Inc. (1988), for 1987 dollars, and have been updated to 1993 dollars using ENR Construction Cost Index. These costs do not include site and surface preparation, and cost of soil cover. Specific cost data for

TABLE 8.10. Installed 1994 Costs for Flexible Membrane Liners.

Material	Thickness mil	Type of Polyester Fabric Reinforcement	Cost $/ft² 1994[b]
CPE	30	—	0.55–0.61
	36	10 × 10–1000d[a]	0.67–0.80
	45		0.80–0.86
CSPE	30	8 × 8–250 d	0.76–0.80
	36	10 × 10–1000 d	0.80–0.86
	45	8 × 8–250d(2)	1.04–1.10
	45	10 × 10–1000 d	0.88–0.95
	60	10 × 10–1000 d	1.35–1.47
HDPE and VLDPE	40	—	0.49–0.61
	60	—	0.67–0.80
	80	—	0.80–0.92
	100	—	0.92–1.10
PVC	30	—	0.33–0.37
	40	—	0.39–0.42
	50	—	0.49–0.55
	60	—	0.61–0.67
PVC-OR	30	—	0.49–0.55
	40	—	0.57–0.64
Nitrile rubber/PVC alloy	30	8 × 8–250d	0.86–0.92
Ethylene interpolymer alloy	30	6.5 oz/yd²	0.86–0.92

[a]d = denier (measure of fineness: 1 d = 1 g/9000 m); oz = ounces. Number in parentheses represents the number of plies of reinforcing fabric.

The original costs were provided by Matrecon, Inc. (1988) in 1987 dollars. These costs have been updated using ENR Construction Cost Indexes (CCI) for 1987 annual average and for March 1994. ENR-CCI for 1987 average = 4400.77; ENR-CCI for March 1994 = 5381.04.

TABLE 8.11. Cost Estimates of Soil Cement, Asphalt Concrete, and Asphalt Membrane Liners.

Liner Type	Installed cost, $/sq yd 1994[a]
Soil cement 6-in thick = sealer (2 coats—each 0.25 gal/sq yd)	11.00
Asphalt concrete, dense-graded paving without sealer coat (hot mix, 4-in thick)	4.16–6.72
Asphalt concrete, hydraulic (hot mix, 4-in thick)	6.87–9.67
Bituminous seal (catalytically blown asphalt) 1 gal/sq yd	3.85
Asphalt emulsion on mat (polypropylene mat sprayed with asphalt emulsion)	1.22–2.44

[a]Estimated installed costs on West Coast. The original costs were provided by Matrecon, Inc. (1988) in 1987 dollars. These costs have been updated using ENR Construction Cost Indexes for 1987 annual average, and for March 1994. ENR-CCI 1987 = 4400.77, ENR-CCI 1994 = 5381.04.
Source: Matrecon, Inc. (1988).

these liners are greatly influenced by geographical location, availability of material, and transportation costs.

8.7 REFERENCES

Acar, Y. B., and A. Ghosh. 1986. "Role of Activity on Hydraulic Conductivity of Compacted Soils Permeate with Acetone," *Proceedings of International Symposium Environmental Geotechnology*, 1:403–412.

Acar, Y. B. and R. K. Seals. 1984. "Clay Barrier Technology for Shallow Land Waste Disposal Facilities," *Hazardous Waste*, 1(2):167–181.

American Society for Testing and Materials (1992), *Annual Book of ASTM Standards*, ASTM, Philadelphia, Pennsylvania (ISBN 0-8031-1754-X).

The Asphalt Institute. 1974. "Asphalt for Wastewater Retention in Fine Sand Areas," The Asphalt Institute Information Series, MISC-74-3, College Park, MD.

The Asphalt Institute. 1976. *Asphalt in Hydraulics*, MS-12, College Park, MD.

Bagchi, A. 1990. *Design, Construction, and Monitoring of Sanitary Landfill*. New York: John Wiley and Sons.

Bingemer, H. and P. Crutzen. 1987. "The Production of Methane from Solid Wastes," *J. Geophysical Research*, 92(D2):2181–2187.

Bright, D. G. 1992. "Polymeric Behavior of Geosynthetic Products," Proceedings of a Seminar on *Structural Geogrids for Waste Containment Applications*, sponsored by Tensar Environmental Systems, Inc., October 21–23, 1992, Morrow, Georgia.

Bureau of Reclamation. 1963. "Linings for Irrigation Canals, Including a Progress

Report on the Lower Cost Canal Lining Program," U.S. Department of the Interior, Washington, D.C.

Chan, P. C., R. Dresnack, J. W. Liskowitz, A. J. Perna and R. Trattner. 1978. "Sorbents for Fluoride Metal Finishing and Petroleum Sludge Leachate Containment Control," EPA-600/2-78-024, U.S. EPA, Cincinnati, OH.

Cheremisinoff, N. P., F. Etterbusch, and A. J. Perna. 1979. *Industrial and Hazardous Wastes Impoundment.* Ann Arbor, Michigan: Ann Arbor Science.

Cronin, D. L., G. L. Freda, J. L. Gray, N. Hausman, C. Pfleiderer and R. W. Watson. 1990. "Operation and Maintenance Considerations for Leachate Collection Systems at a Sanitary Landfill Facility in New Jersey," Proceedings of the 1990 WPCF National Specialty Conference on *Water Quality Management of Landfills,* Chicago, IL, July 15-18, pp. 7-68-7-77.

Curran, M. A. and R. K. Frobel. 1985. "Strength and Durability of Flexible Membrane Liner Seams after Short-Term Exposure to Selected Chemical Solutions," *Land Disposal of Hazardous Wastes,* Proceedings of Eleventh Annual Research Symposium, U.S. Environmental Protection Agency, EPA/600/9-85/013, Cincinnati, Ohio, pp. 307-312.

Daniels, D. 1988. "Clay Liners," Seminars – Requirements for Hazardous Waste Landfill Design, Construction and Closure, CERI-88-33, *U.S. EPA,* Technology Transfer, Cincinnati, Ohio, pp. II-1–II-47.

Demetrocopoulos, A. M., G. P. Korfiatis, E. L. Bourodimos, and E. G. Nawy. 1984. "Modeling for Design of Landfill Bottom Liners," *Journal of the Environmental Engineering Division, ASCE,* 110(EE6):1084-1098.

E. C. Jordan Company. 1984. "Performance Standard for Evaluating Leak Detection," Draft Final Report, U.S. EPA Contract N. 68-01-6871, Work Assignment No. 32, US EPA, Washington, D.C.

Ewald, G. W. 1973. "Stretching the Life of Synthetic Pond Liners," *Chemical Engineering,* 80(22):67-69.

Goldman, L. J., L. I. Greenfield, A. S. Damle, G. L. Kingsbury, C. M. Northeim, and R. S. Truesdale. 1990. *Clay Liners for Waste Management Facilities: Design, Construction, and Evaluation.* Park Ridge, NJ: Noyes Data Corporation.

Haxo, H. E. and R. White 1976. "Evaluation of Liner Materials Exposed to Leachate," Second Interim Report, U.S. EPA-600/2-76-255, U.S. EPA, Cincinnati, Ohio.

Haxo, H. E. 1977. "Compatibility of Liners with Leachate," Proceedings of the *Third Annual Municipal Solid Waste Research Symposium,* EPA-600/2-77-081, U.S. EPA, Cincinnati, Ohio, pp. 149-159.

Haxo, H. E. 1978. "Interaction of Selected Lining Materials with Various Hazardous Wastes," Proceedings of the 4th Annual Research Symposium. EPA-600/9-78-016, U.S. EPA, Cincinnati, Ohio, pp. 256-272.

Haxo, H. E., et al. 1985. *Liner Material for Hazardous and Toxic Wastes and Municipal Solid Waste Leachate.* Park Ridge, NJ: Noyes Publications.

Haxo, H. E. and T. P. Caney. 1988. "Transport of Dissolved Organics from Dilute Aqueous Solutions Through Flexible Membrane Liners," *Hazardous Waste and Hazardous Materials,* 5(4):275-294.

Heglie, J., T. Bray, D. Bostrom, and G. Wolff. 1990. "Planning for Leachate Management at a Closed South Bay Landfill," Proceedings of the 1990 WPCF National Specialty Conference on *Water Quality Management of Landfills,* Chicago, IL. July 15-18, pp. 7-1-7-17.

Kmet, P., K. J. Quinn, and C. Slavic. 1981. "Analysis of Design Parameters Affecting the Collection Efficiency of Clay Liner Landfills," Proceedings of Fourth Annual Madison Conference of Applied Research and Practice on *Municipal and Industrial Waste,* University of Wisconsin, Madison, Wisconsin.

Koerner, R. E. 1984. *Designing with Geosynthetics,* Second Edition. Englewood Cliffs, NJ: Prentice Hall.

Koerner, R. E., Y. Halse-Hsuan, and A. R. Lord. 1991. "Long-Term Durability of Geomembranes," *Civil Engineering,* ASCE, pp. 6–58.

Korfiatis, G. P. and A. M. Demetrocopoulos. 1986. "Flow Characteristics of Landfill Leachate Collection Systems and Liners," *Journal of the Environmental Engineering Division, ASCE,* 112(3):538–550.

Kumar, J. and J. A. Jedlicka. 1973. "Selecting and Installing Synthetic Pond Linings," *Chemical Engineering,* 80(5):805–820.

Lentz, J. J. 1981. "Apportionment of Net Recharge in Landfill Covering Layer into Separate Components of Vertical Leakage and Horizontal Seepage," *Water Resources Research,* 17(4):1231–1234.

Loehr, R. C. 1987. "Land Disposal of Hazardous Wastes," in *Hazardous Waste Management Engineering,* E. G. Martin and J. H. Johnson, eds. New York: Van Nostrand Reinhold Co.

Lu, J. C. S., B. Eichenberger, and R. J. Stearns. 1984. "Production and Management of Leachate from Municipal Landfills: Summary and Assessment," U.S. Environmental Protection Agency EPA 600/2-84-092, Municipal Environmental Research Laboratory, Cincinnati, Ohio.

Lu, J. C. S., B. Eichenberger, and R. J. Stearns. 1985. *Leachate from Municipal Landfills: Production, and Management.* Park Ridge, New Jersey: Noyes Publishers.

Lutton, R. J., G. L. Regan, and L. W. Jones. 1979. "Design and Construction of Covers for Solid Waste Landfills," EPA-600/2-79-165, U.S. Environmental Protection Agency, Municipal Environmental Research Laboratory, Cincinnati, Ohio.

Matrecon, Inc. 1980. "Lining of Waste Impoundment and Disposal Facilities," SW-870, U.S. Environmental Protection Agency, Cincinnati, Ohio.

Matrecon, Inc. 1988. "Lining of Waste Impoundment and Other Impoundment Facilities," USEPA/600/2-88/052, Risk Reduction Engineering Laboratory, Office of Research and Development, U.S. Environmental Protection Agency, Cincinnati, Ohio.

McBean, E. A., R. Poland, and F. A. Rovers. 1982. "Leachate Collection Design for Containment Landfills," *Journal of the Environmental Engineering Division, ASCE,* 108(EE1):204–209.

McEnroe, B. M. and P. R. Schroeder. 1988. "Leachate Collection in Landfills: Steady Case," *Journal of the Environmental Engineering, ASCE,* 114(EE5):1052–1062.

Morrison, W. R. and L. D. Parkhill. 1987. "Evaluation of Flexible Membrane Liner Seams After Chemical Exposure and Simulated Weathering," U.S. Environmental Protection Agency, Hazardous Waste Engineering Research Laboratory, EPA/600/S2-87/015, Cincinnati, Ohio.

Poly-Flex, Inc. (undated). "An Engineering Approach to Groundwater Protection," Reference Manual, Grand Prairie, Texas.

Sai, J. and J. D. Zabcik. 1985. "Estimate of Surface Impoundment Construction Costs Under the RCRA Amendments of 1984," Contract No. 68-03-1816, U.S. EPA, Cincinnati, Ohio.

Styron, C. R. and Z. B. Fry. 1979. "Flue Gas Cleaning Sludge Leachate/Liner Compatibility Investigation—Interim Report," EPA-600/2-79-136, PB 80-100480 U.S. EPA, Cincinnati, Ohio.

Tchobanoglous, G., H. Theisen, and S. Vigil. 1993. *Integrated Solid Waste Management Issues, second edition,* New York, McGraw-Hill.

U.S. Congress, Office of Technology Assessment. 1989. "Facing Americas' Trash: What Next for Municipal Solid Waste," OTA-424, U.S. Government Printing Office.

U.S. Environmental Protection Agency. 1983. "Lining of Waste Impoundment and Disposal Facilities," SW-870, Municipal Environmental Research Laboratory, Cincinnati, Ohio.

U.S. Environmental Protection Agency. 1986. "Technical Guidance Document, Construction Quality Assurance for Hazardous Waste Disposal Facilities," EPA/530-SW-86-031, Cincinnati, Ohio.

U.S. Environmental Protection Agency. 1988. Office of Solid Waste, Design Criteria (Subpart D), "Criteria for Municipal Solid Waste Landfills," EPA/530-SW-88-042, Washington, D.C.

Wong, J. 1977. "The Design of a System for Collecting Leachate from a Lined Landfill Site," *Water Resources Research,* 13(2):404–409.

Woods, R. 1992. "Building a Better Liner System," *Waste Age,* 23(3):26–32.

CHAPTER 9

Leachate Treatment

9.1 INTRODUCTION

LANDFILL leachate initially is a high-strength wastewater, characterized by low pH, high biochemical oxygen demand (BOD), chemical oxygen demand (COD), and by the presence of toxic chemicals. In addition, the leachate quality is variable from landfill to landfill, and over time as a particular landfill ages. Consequently, neither conventional biological waste treatment nor chemical treatment processes separately achieve high removal efficiency over the life of the fill.

Selection and design of a leachate treatment process is not simple. Important factors that govern the selection and design of treatment facilities include leachate characteristics, effluent discharge alternatives, technological alternatives, costs, and permit requirements. The purpose of this chapter is to review the treatability and nature of leachate treatment problems, various treatment methods, their application and limitations, and finally to develop some generalized methods and process trains for leachate treatment under different conditions.

9.2 LEACHATE TREATMENT PROBLEMS

The task of planning and designing a facility for treatment of leachate from sanitary landfills requires knowledge of the landfill design, leachate quantity and quality, degree of treatment needed, and ultimate disposal methods of effluent and residues. Many nontechnical factors, such as legal issues, regulatory constraints and public participation, may also influence the planning and design. Furthermore, most of the facilities are designed to provide service over the landfill life expectancy (30 years or more). During this extended time span, leachate quality will change, technology may improve, new laws may be passed, new regulations may be issued, and economic factors may change. The engineer must therefore consider these possibilities, and should favor treatment processes that are sufficiently flexible to remain useful, even with changing technology, regulations, leachate characteristics, and economics.

Specific problems inherent with treatment of landfill leachate are:

(1) The high strength of waste and magnitude of pollution potential dictates the selection and use of reliable treatment processes.

(2) The changes encountered from landfill to landfill are such that waste treatment techniques applicable at one site may not be directly transferable to other locations. It may be necessary that each instance be separately engineered for proper treatment.

(3) The source of leachate is primarily percolating water that may be seasonal depending on hydrologic and climatic factors.

(4) The chemical nature of the solid wastes accepted at a landfill has a marked effect on the composition of the leachate.

(5) The fluctuations in the leachate quantity and quality, which occur over both short and long time intervals, must be considered in the treatment plant design. The process designed to efficiently treat the leachates from a young landfill should be modified in the future to treat the leachate adequately as the landfill ages, or effluent standards change.

9.3 REVIEW OF APPLICABLE WASTEWATER TREATMENT PROCESSES

Once leachate from a sanitary landfill is collected, numerous alternatives exist for treatment and disposal. If the facility is not in operation, the quantity and characteristics of the leachate to be treated must be determined. Other factors determining the type and degree of treatment are:

(1) Leachate characteristics—organic and inorganic concentrations

(2) Hazardous nature—high concentrations of organic and inorganic toxic chemicals

(3) Discharge alternatives—surface waters, publicly owned treatment works, land treatment, effluent used on landfill site

(4) Degree of treatment—leachate characteristics, permit requirement, discharge alternatives

(5) Treatability studies—available experimental data, and applicable technologies.

(6) Operational needs—analytical testing, personnel safety training, equipment repairs and maintenance

(7) Costs—availability of funds, post closure requirements

The leachate characteristics depend on the nature of the wastes landfilled, and on the stage of decomposition in the fill. If the characteristics of the collected leachate indicate it is a hazardous waste, the leachate must be managed as a hazardous waste in accordance with the applicable permit requirements. If the leachate is discharged as a point source to the surface waters, it is subject to point source permit limitations. If the leachate is discharged to a publicly owned treatment works, pretreatment requirements must apply. The applicable methods of leachate treatment are biological, physical and chemical, a combination of these processes, and combined municipal wastewater treatment. Many physical, chemical, and biological treatment processes applicable to landfill leachate treatment, along with brief descriptions, are provided in Table 9.1. Readers are referred to several excellent textbooks for theory and design of these processes (Tchobanoglous and Schroeder 1985; Metcalf and Eddy 1991; Qasim 1994). The process discussions in these sources are along the traditional lines, mainly applicable to treatment of municipal wastewater. For treatment of landfill leachate, the suggested general approach is to utilize physical treatment processes in conjunction with (a) biological treatment and (b) chemical treatment, in particular, coagulation-flocculation, precipitation, chemical oxidation, carbon adsorption and membrane separation processes. A general discussion is presented below to provide the readers with necessary background information on these processes as they apply to leachate treatment.

9.3.1 BIOLOGICAL WASTE TREATMENT

Biological treatment processes involve placing a waste stream in contact with a mixed culture of microorganisms. The microorganisms consume the organic matter in the waste stream. The process must provide optimum environmental conditions for enhanced degradation of organic wastes. The methods for optimizing the biological degradation include controlling the dissolved oxygen level, adding nutrients, increasing the concentration of microorganisms, and maintaining many environmental factors such as pH, temperature, and mixing. In aerobic biological treatment processes, the organics are decomposed to carbon dioxide and water. Oxygen is essential for decomposition of organic matter. In anaerobic treatment processes, organics are decomposed in the absence of free oxygen. Methane and carbon dioxide are the major end products. The biological treatment processes do not alter or destroy inorganics. Trace concentrations of inorganics may be partially removed from the liquid waste stream during the biological treatment because of precipitation and adsorption onto the microbial cells. Typically, microorganisms have a net negative charge and are therefore able to perform cation exchange with metal ions in solution. Anionic spe-

TABLE 9.1. Physical, Chemical, and Biological Treatment Processes Applicable to Process Trains for Leachate Treatment.

Process	Description
I. Physical	
A. Equalization	Flow and mass loadings are equalized by means of utilizing in-line or off-line equilization chambers.
B. Screening	Suspended and floating debris are removed. Removal is by straining action.
C. Flocculation	Fine particles are aggregated. Gentle stirring is utilized.
D. Sedimentation	Settleable solids and floc are removed by gravity.
E. Flotation	Solids are floated by fine air bubbles and skimmed from the surface. Dissolved air flotation (DAF) is commonly used.
F. Air stripping	Air and liquid are contacted in countercurrent flow in a stripping tower. Ammonia, other gases and volatile organics are removed.
G. Filtration	Suspended solids and turbidity are removed in a filter bed or microscreen.
H. Membrane processes	These are demineralization processes. Dissolved solids are removed by membrane separation. Ultrafiltration, reverse osmosis and electrodialysis are the most common systems.
I. Natural evaporation	The waste is impounded in basins that have an impervious liner. Liquid is evaporated. The rate of evaporation depends upon temperature, wind velocity, humidity, and natural precipitation.

TABLE 9.1. (continued).

Process	Description
II. Chemical	
A. Coagulation	Colloidal particles are destabilized by rapid dispersion of chemicals. Organics, suspended solids, phosphorus, some metals, and turbidity are removed. Alum, iron salts and polymers are commonly used coagulation chemicals.
B. Precipitation	Solubility is reduced by chemical reaction. Hardness, phosphorus, and many heavy metals are removed.
C. Gas transfer	Gases are added or removed by mixing, air diffusion, and change in pressure.
D. Chemical oxidation	Oxidizing chemicals such as chlorine, ozone, potassium permanganate, hydrogen peroxide and oxygen are used to oxidize organics, hydrogen sulfide, ferrous, and other metal ions. Ammonia and cyanide are oxidized by strong oxidizing chemicals.
E. Chemical reduction	Metal ions are reduced for precipitation, recovery, and conversion into a less toxic state (chromium). Many metals are also removed. Oxidizing chemicals are reduced (dechlorination). Common reducing chemicals are sulfur dioxide, sodium bisulfite, and ferrous sulfate.
F. Disinfection	Destruction of pathogens is achieved by using oxidizing chemicals, or ultraviolet light.
G. Ion exchange	Removal of inorganic species is achieved from liquid. Ammonia is selectively removed by *clinoptilite* resin. This process is used for demineralization.
H. Carbon adsorption	Used for reduction of residual BOD, COD, toxic and refractory organics. Some heavy metals are also removed. Carbon is used in powdered form, or in a granular bed.

(continued)

TABLE 9.1. (continued).

Process	Description
III. Biological processes	Microorganisms are cultivated to consume biodegradable organic matter. Biological processes are also used for nitrification and denitrification, and enhanced phosphorus removal.
A. Aerobic	Microorganisms are cultivated in the presence of molecular oxygen. Solids are recirculated. The end product is carbon dioxide.
1. Suspended growth	The wastewater containing BOD, solids, and nutrients are mixed with a large population of active microorganisms suspended in an aeration basin.
a. Activated sludge	In the activated sludge process the food and sludge microorganisms are aerated. The microorganisms are settled and recirculated. Common process modifications are conventional, tapered aeration, step aeration, completely mixed, pure oxygen, extended aeration, and contact stabilization.
b. Nitrification	Ammonia nitrogen is oxidized to nitrate. BOD removal can also be achieved in a single aeration basin, or in a separate basin.
c. Aerated lagoon	Large aeration basins with several days of detention period are used.
d. Sequencing Batch Reactor (SBR)	A SBR is a fill-and-draw activated sludge treatment system. Food and microorganism contact, organics stabilization, sedimentation and discharge of clarified effluent occur in a single basin.
2. Attached growth	The population of active microorganisms is supported over solid media. The solid media may be of rocks or synthetic material.
a. Trickling Filters	Water is applied over a bed of rocks or synthetic media. Trickling filters are slow rate, high rate, super rate or roughing, and two stage filters. Aeration is by natural draft or forced draft.
b. Rotating Biological Contactor (RBC)	Consists of a series of closely spaced circular contactor disks of synthetic material. The disks are partly submerged in the wastewater.

TABLE 9.1. (continued).

Process	Description
3. Combined suspended and attached growth	The system has microorganisms in suspension and attached to a solid media. The process effectively removes BOD, total suspended solids, and achieves nitrification. It is extensively used for treatment of high strength industrial waste streams.
B. Anaerobic	The microorganisms are cultivated in the absence of oxygen. The complex organics are solubilized and stabilized. Carbon dioxide, methane and other organic compounds are the end products.
1. Suspended growth	The waste is mixed with biological solids in a digester, and the contents are commonly stirred, and heated to an optimum temperature.
a. Conventional	High organic strength waste or sludge is stabilized in a digester. The digesters are standard rate, high rate, one-stage, or two-stage.
b. Contact process	The waste is digested in a completely mixed anaerobic reactor. The digested solids are settled in a clarifier and then returned to the digester.
c. Upflow Anaerobic Sludge Blanket (UASB)	Waste enters the bottom and flows upward through a blanket of biologically formed granules or solids.
d. Denitrification	Nitrite and nitrate are reduced to gaseous nitrogen in an anaerobic environment. A suitable organic carbon source (acetic acid, methanol, sugar etc.) is required.
e. Combined anoxic, anaerobic and aerobic system	Nitrogen and phosphorus are removed along with BOD in an anoxic, anaerobic and aerobic treatment system. Nitrate is converted to gaseous nitrogen in the anoxic reactor. Phosphorus is released in anoxic and anaerobic reactors. Uptake of released phosphorus, BOD stabilization, and nitrification of ammonia occurs in the aerobic reactor.

(continued)

TABLE 9.1. (continued).

Process	Description
(Biological processes, anaerobic, *continued*)	
2. Attached growth	The microbiological film is supported over a solid media. The organic matter is stabilized as the waste comes in contact with the attached growth.
a. Anaerobic filter	The reactor is filled with the solid media, and the waste flows upward. Medium-strength wastes are treated in a relatively short hydraulic retention time.
b. Expanded bed or fluidized bed	The reactor is filled with media such as sand, coal, and gravel. The influent and recycled effluent are pumped from the bottom. The bed is kept in an expanded condition. This process has been used to dilute wastes.
c. Rotating biodisks	Circular disks are mounted on a central shaft and rotated while completely submerged in an enclosed housing. Biofilm grows over the disks and stabilizes the organic wastes.
d. Denitrification	The attached growth in an anaerobic environment, and in the presence of a carbon source, reduces nitrite and nitrate into gaseous nitrogen.
3. Combined suspended and attached growth	The attached and suspended microbiological growth in an anaerobic environment consumes the organic matter.
C. Aerobic-anaerobic stabilization ponds	Stabilization ponds are earthen basins with impervious liner. The basins may be aerobic, facultative, or anaerobic depending upon the depth and strength of wastes. Source of oxygen is natural aeration.
D. Land treatment	The waste is applied over land to utilize plants, soil matrix and natural phenomena to treat waste by a combination of physical, chemical, and biological means. The methods of land application are slow-rate irrigation, rapid infiltration-percolation, and over-land flow.

cies, such as chlorides and sulfates, are not affected by biological treatment.

With a properly acclimatized microbial population, and adequate equalization as preliminary treatment to ensure a uniform hydraulic flow and organic concentration, biological treatment is applicable to many industrial wastes. For the treatment of organics in an aqueous medium, it is probably the most cost-effective treatment. Energy and chemical demands are low compared to other processes; however, land requirements are greater. Several of the most common waste treatment processes that have been applied to leachate treatment are:

- Activated sludge
- Waste stabilization pond
- Aerated lagoon
- Trickling filter
- Rotating biological contactor
- Anaerobic digestion

An overview and performance of these processes is provided in Table 9.2 (Berkowitz et al. 1978a, 1978b, 1978c).

9.3.1.1 Activated Sludge

In the activated sludge process, microorganisms (MO) are mixed thoroughly with the organics so that they can grow and stabilize the substrate. As the microorganisms grow in the presence of oxygen and are mixed by agitation, the individual organisms flocculate to form an active mass of microbial floc called activated sludge. The mixture of the activated sludge and wastewater in the aeration basin is called mixed liquor suspended solids (MLSS), and the volatile organic fraction is called mixed liquor volatile suspended solids (MLVSS). The MLSS flows from the aeration basin into a secondary clarifier where the activated sludge is settled. A portion of the settled sludge is returned to the aeration basin to maintain the proper food-to-microorganism (MO) ratio to permit rapid breakdown of the organic matter. Because more activated sludge is produced than that can be used in the process, some of it is wasted from the aeration basin or from the return sludge line to the sludge-handling systems, for further treatment and disposal. Air is introduced into the aeration basin, either by diffusers or by mechanical mixers. There are many modifications of the activated sludge process. These modifications differ in mixing and flow pattern in the aeration basin, and in the manner in which the microorganisms are mixed with the incoming wastewater. The modifications and design parameters are summarized in Table 9.3. The process diagrams are shown in Figure 9.1. The principal types of aeration basins are plug flow,

TABLE 9.2. Overview of Biological Treatment Processes.

Biological Process	Principal Microbial Population	Optimum Temp.	Range in pH	% Solid in Waste Stream	Average Retention Time, day	Organics Decomposed	Upper Limit of BOD Effectively Handled, mg/L	Effluent	Residue
Activated sludge	Aerobic and heterotrophic bacteria	Mesophilic	6–8	<1%	<1	All but oil, and halogenated aromatics	<10,000	CO_2 and 5–15% influent BOD remains	Biomass sludge
Waste stabilization	Aerobic, anaerobic, faculative heterotrophic bacteria, and autotrophic algae	Mesophilic	6–8	<0.1%	10–120	Mostly carbohydrates, proteins, organic acids and alcohols	<1000	CO_2 and 10–40% influent BOD remains	None
Aerated lagoon	Aerobic, heterotrophic and facultative bacteria	Mesophilic	6–8	<1%	2–7	All but oil, and halogenated aromatics	<5,000	CO_2 and 10–30% influent BOD remains	Biomass sludge
Trickling filter	Aerobic and heterotrophic bacteria	Mesophilic	6–8	<1%	<1	All but oil, grease and halogenated aromatics, nitrogen compound	<5,000	CO_2 and 10–30% influent BOD remains	Biomass sludge

TABLE 9.2. (continued).

Biological Process	Principal Microbial Population	Optimum Temp.	Range in pH	% Solid in Waste Stream	Average Retention Time, day	Organics Decomposed	Upper Limit of BOD Effectively Handled, mg/L	Effluent	Residue
Rotating biological contactor	Aerobic and heterotrophic bacteria	Mesophilic	6–8	<1%	<1	All but oil, halogenated aromatics nitrogen compound	<5000	CO_2 and 10–20% influent BOD remains	Biomass sludge
Anaerobic digester	Obligate anaerobic and hetero-trophic bacteria	Thermophilic	6.4–7.5	<10%	10–30	Carbohydrates, proteins, organic acids and alcohols	Not applicable	CO_2 and CH_4, 40–60% influent volatile solids remain	Stabilized sludge

Source: Adapted from Berkowitz et al. (1978a, 1978b, and 1978c).

TABLE 9.3. Description and Design Parameters of Various Activated Sludge Process Modifications.

Process Modification	Brief Description	Flow Regime	Sludge Retention Time, θ_c(d)	Food to MO Ratio[a](d^{-1})	Aerator Loading[b] (kg/m³·d)	MLSS,[c] (mg/L)	Aeration Period (h)	Recirculation Ratio Q_r/Q
Conventional	The influent and returned sludge enter the tank at the head end of the basin and are mixed by the aeration system [Figure 9.1(a)].	Plug	5–15	0.2–0.4	0.3–0.6	1500–3000	4–8	0.25–0.5
Tapered aeration	The tapered aeration system is similar to the conventional activated sludge process. The major difference is in arrangement of the diffusers. The diffusers are close together at the influent end where more oxygen is needed. The spacing of diffusers is increased toward the other end of the aeration basin.	Plug	5–15	0.2–0.4	0.3–0.6	1500–3000	4–8	0.25–0.5

TABLE 9.3. (continued).

Process Modification	Brief Description	Flow Regime	Sludge Retention Time, θ_c(d)	Food to MO Ratio^a(d^{-1})	Aerator Loading^b (kg/m^3·d)	MLSS,^{c} (mg/L)	Aeration Period (h)	Recirculation Ratio Q_r/Q
Step aeration	The settled wastewater is applied at several points in the aeration basin. Generally, the tank is subdivided into three or more parallel channels with around-the-end baffles. The oxygen demand is uniformly distributed [Figure 9.1(b)].	Plug	5–15	0.2–0.4	0.6–1.0	2000–3000	3–5	0.25–0.75
Completely mixed aeration	The influent and the returned sludge are mixed and applied at several points along the length and width of the basin. The contents are mixed and the MLSS flows across the tank to the effluent channel. The oxygen demand and organic loading are uniform along the entire length of the basin [Figure 9.1(c)].	Completely mixed	5–15	0.2–0.6	0.8–2.0	3000–6000	3–5	0.25–1.00

(continued)

TABLE 9.3. (continued).

Process Modification	Brief Description	Flow Regime	Sludge Retention Time, θ_c(d)	Food to MO Ratio[a](d^{-1})	Aerator Loading[b] (kg/m³·d)	MLSS,[c] (mg/L)	Aeration Period (h)	Recirculation Ratio Q_r/Q
Extended aeration	The extended aeration process utilizes large aeration basin where high population of MO is maintained. It is used for small flows. Prefabricated package plants utilize this process extensively. *Oxidation ditch* is a variation of extended aeration process. It has channel in shape of a race track. Rotors are used to supply oxygen and maintain circulation [Figure 9.1(d)].	Completely mixed or plug	20–30	0.05–0.15	0.1–0.4	3000–6000	18–36	0.5–1.0
Contact stabilization	The activated sludge is mixed with influent in the contact tank in which the organics are absorbed by MO. The MLSS is settled in the clarifier. The returned sludge is aerated in the reaeration basin to stabilize the organics. The process requires approximately 50 percent less tank volume [Figure 9.1(e)].	Plug	5–15	0.2–0.6	1.0–1.2	1000–4000[d] 4000–10,000[e]	0.5–1.0[d] 3.0–6.0[e]	0.5–1.0

TABLE 9.3. (continued).

Process Modification	Brief Description	Flow Regime	Sludge Retention Time, θ_c(d)	Food to MO Ratio[a](d^{-1})	Aerator Loading[b] (kg/m³·d)	MLSS,[c] (mg/L)	Aeration Period (h)	Recirculation Ratio Q$_r$/Q
Pure oxygen	Oxygen is diffused into covered aeration tanks. A portion of gas is wasted from the tank to reduce the concentration of CO_2. The process is suitable for highstrength wastes where space may be limited. Special equipment for generation of oxygen is needed [Figure 9.1(f)].	Completely mix	8–20	0.25–1.0	1.6–3.3	6000–8000	2–5	0.25–0.5

[a]Food to microorganism ratio (F/M) is kg BOD$_5$ applied per day per kg of MLVSS in the aeration basin.
[b]Aerator loading is kg of BOD$_5$ applied per day per cubic meter of aeration capacity.
[c]Generally the ratio of MLVSS to MLSS is 0.75–0.85.
[d]Contact tank.
[e]Reaeration or stabilization tank.
Source: Adopted from Qasim (1994).

229

completely mixed, and arbitrary flow reactors. The design of an activated sludge process utilizes empirical as well as rational parameters based on biological kinetic equations. These equations express biological (sludge) growth and substrate utilization rates in terms of biological kinetic coefficients. Using these equations, the design parameters such as volume of aeration basin, effluent quality, rates of return sludge and waste sludge, aeration period, oxygen utilization rates, food-to-MO ratio, and mean cell

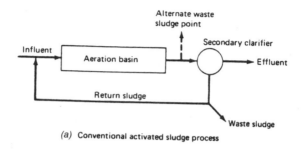

(a) Conventional activated sludge process

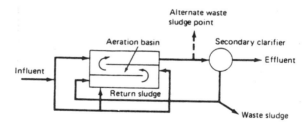

(b) Step-aeration activated sludge process

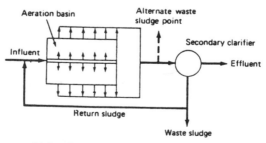

(c) Completely-mixed activated sludge process

Figure 9.1 Modifications of activated sludge process. Source: adapted from Qasim (1994).

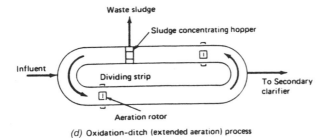

(d) Oxidation-ditch (extended aeration) process

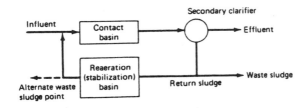

(e) Contact stabilization process

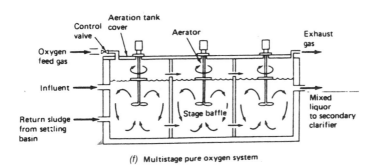

(f) Multistage pure oxygen system

Figure 9.1 (continued).

residence time can be calculated. Details on biological kinetic models and the derivation procedure for these equations may be found in several textbooks (Tchobanoglous and Shroeder 1985; Metcalf and Eddy 1991; Qasim 1994).

The values of kinetic coefficients Y, k, k_d, and K_s (Table 9.4) greatly influence the design of the activated sludge process. These values depend on the characteristics of the wastewater and therefore must be determined for

TABLE 9.4. Typical Values of Kinetic Coefficients for
Activated Sludge Process Treating Municipal Wastewater.

Kinetic Coefficient	Definition	Basis	Values	
			Range	Typical
k	Maximum rate of substrate utilization per unit mass of MO	d^{-1}	2–8	4
k_d	Endogenous decay coefficient	d^{-1}	0.03–0.07	0.05
K_s	Substrate concentration at one-half of the maximum growth rate	mg/L, BOD_5 mg/L, COD	40–120 20–80	80 40
Y	Yield coefficient over infinite period of log growth	VSS/BOD_5 VSS/COD	0.3–0.7 0.2–0.5	0.5 0.4

Source: Adapted from Qasim (1994).

every waste stream (especially if it contains industrial wastes) from bench or pilot plant studies. The range and typical values of these kinetic coefficients for municipal wastewater are summarized in Table 9.4.

9.3.1.2 Waste Stabilization Pond

Waste stabilization ponds are large, shallow basins where wind action provides aeration, and a mixed autotrophic and heterotrophic microbial population provides decomposition of organics over a long retention time. Deep ponds, more than 1.5 m deep, may promote anaerobic decomposition of settled sludge. These ponds are a popular means of wastewater treatment for small communities, and for industries that produce organic waste streams. Stabilization ponds have the advantage of low construction and operation costs. The major disadvantages are large land area requirements, odor and insect problems, possible groundwater contamination, and poor effluent quality.

In a stabilization pond, solids settle to the bottom. A wide variety of microscopic plants and animals find the environment a suitable habitat. Organic matter is metabolized by bacteria and protozoa as primary feeders. Secondary feeders include protozoa and higher animals such as rotifers and crustaceans. The nutrients released are utilized by algae and other aquatic plants. The main sources of oxygen are natural reaeration and photosynthesis. In the bottom layer, the accumulated solids are actively decomposed by anaerobic bacteria. Stabilization ponds are usually classified as *aerobic*, *facultative*, and *anaerobic*. This classification is based on the nature of the biological activity taking place. Design factors such as depth, detention time, organic loading, and effluent quality also vary greatly for the three types of lagoons. These values are provided in Table 9.5. Waste stabilization ponds have been widely used for sanitary sewage and dilute industrial wastes, mostly to provide final effluent polishing.

9.3.1.3 Aerated Lagoon

The aerated lagoon technique was developed by adding artificial aeration to existing waste stabilization ponds. Usually, the lagoon is an earthen or lined basin with sloping sides and about 2 to 5 m deep. For the treatment of industrial wastes, it may be necessary to line the basin with an impermeable material. Because wastewaters in aerated lagoons are generally not as well mixed as those in the activated sludge basins, a portion of the solids may settle to the bottom and undergo anaerobic microbial decomposition. Retention times are several days. The process has been successfully used for petrochemical, textile, pulp and paper mill, cannery and refinery wastewaters. The aerated lagoon employs essentially the same microbial reactions as activated sludge process. There is, however, no sludge recycle system for continuous circulation of microorganisms, and microbial strains do not acclimatize to the same extent. BOD removal efficiencies range from 60 to 90%. The process is not as attractive for industrial waste treatment as the activated sludge process. The removal efficiencies are not as high and the process is less flexible in maintaining effluent limitations under varied influent loading. In the absence of a clarifier the suspended solids concentration in the aerated lagoon and in the effluent is in the range of 80–250 mg/L. Chemical coagulation and clarification of effluent from aerated lagoons will produce well nitrified effluent.

9.3.1.4 Trickling Filter

In a trickling filter the wastewater is sprayed through the air to absorb oxygen, and then allowed to trickle through a bed of rock or synthetic media coated with a slime of microbial growth. Process modifications employ various media and depths to retain the microorganisms under varying hydraulic and effluent recycle conditions. Primarily, the metabolic process is aerobic and the microbial population is similar to that of the activated sludge process. However, because of the relatively short contact time, the percentage removal of organics is not as large. The process involves open tanks or towers to house the filter packing, followed by effluent clarifiers. Recycle pumps may be used to recirculate the filter effluent. A rotating spray dosing system distributes the influent wastewater to the filter surface.

TABLE 9.5. Process Description and Design Parameters for Stabilization Ponds.

Parameter	Aerobic	Facultative (Aerobic-Anaerobic)	Anaerobic
Process description	Aerobic condition is maintained throughout entire depth	Upper layer aerobic, middle layer aerobic-anaerobic, bottom layer anaerobic	Anaerobic condition throughout the pond
Detention time, d	5–20	10–30	20–50
Water depth, m	0.3–1[a]	1–2	2.5–5
BOD_5 loading, kg/ha · d	40–120[b]	15–120	200–500
BOB_5 removal, %	40–80[c]	70–90	60–90
TSS mainly due to algae, mg/L	100–250	40–100	70–120

[a]1 m = 3.28 ft.
[b]1 kg/ha · d = 0.8922 lb/acre·d.
[c]Overall BOD_5 removal is small because of high concentration of algae in the effluent.
Source: Adapted from Qasim (1994).

Trickling filters have been extensively used in sewage treatment and in treatment of refinery wastewaters containing oil, phenol and sulfide. Like activated sludge process, the trickling filters are also applicable to industrial waste streams. BOD removal efficiencies range from 60 to 85%. Energy demands are low, almost one-tenth that of activated sludge treatment. The process could be used in sequence with other biological treatment, such as activated sludge, but it is not generally efficient enough to be used as the sole method of biodegradation. Based on the organic and hydraulic loading, trickling filters are classified as *low-rate, intermediate-rate, high-rate,* and *super-rate* (roughing filters). Often two-stage trickling filters (two trickling filters in series) are used for treating high-strength wastes. Typical design information for trickling filters are summarized in Table 9.6.

9.3.1.5 Rotating Biological Contactor (RBC)

A rotating biological contactor (also called bio-disk process) consists of a series of circular plastic plates (disks) mounted over a shaft that rotates slowly. These disks remain approximately 40 percent immersed in a contoured bottom tank. The disks are spaced so that wastewater and air can reach the exposed surface. The biological growth develops over the disks, which receive alternating exposures to organics and the air. The excess growth of the microorganisms becomes detached and therefore the effluent requires clarification. The rotating biological contactor has low power demand, greater process stability, higher organic loadings, and smaller quantity of waste sludge than the activated sludge process. The rotating biological contactor is shown in Figure 9.2.

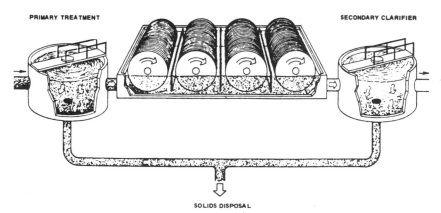

Figure 9.2 Flow scheme of a rotating biological contactor. Source: adapted from Culp and Heim (1978).

TABLE 9.6. Typical Design Information for Different Types of Trickling Filters.

Item	Low Rate	Intermediate	High Rate	Super Rate or Roughing	Two-stage
Operation	Intermittent	Continuous	Continuous	Continuous	Continuous
Recirculation ratio	0	0–1.0	1.0–2.5	1.0–4.0	0.5–3.0
Depth, m	1.5–3.0	1.25–2.5	1.0–2.0	4.5–12	2.0–3.0
Hydraulic loading, $m^3/m^2 \cdot d$	1–4	4–10	10–40	40–200	10–40
BOD_5 loading, $kg/m^3 \cdot d$	0.08–0.32	0.24–0.48	0.32–1.0	0.08–6.0	1.0–2.0
Sloughing	Intermittent	Intermittent	Continuous	Continuous	Continuous
Media	Rock, slag	Rock, slag	Rock, slag, synthetic	Synthetic	Rock, slag, synthetic
Filter flies	Many	Medium	Small	None	None
Power $kW/10^3 m$	2–4	2–8	6–10	10–20	6–10
BOD_5 removal efficiency, percent	74–80	80–85	80–85	60–80	85–95
Effluent	Well nitrified	Well nitrified	Little nitrification	Little nitrification	Well nitrified

Source: Adapted from Qasim (1994).

9.3.1.6 Anaerobic Digestion

The high strength organic waste or sludge is processed in an anaerobic digester to stabilize organics and improve stability. In a conventional process, the waste stream is fed into the middle zone of an air tight closed tank, with or without an agitating mechanism. The solids are digested by microorganisms, and gases rise bringing the scum to the surface. The gases are collected from the roof of the tank and used to maintain the temperature of the digester and sometimes to heat other parts of the plant. The anaerobic bacteria are sensitive to pH, temperature, and the composition of the waste, but usually with time the system will achieve its own balance. Modifications have been made for different installations, usually involving agitation, heating, and maintaining high biological solids, which may shorten the process by several days. Most installations have been used for the digestion of sewage sludge. Also, in the past 20 years, the system has been studied for many high and medium strength industrial wastes.

Anaerobic digestion involves a complex biochemical process in which several groups of facultative and anaerobic organisms simultaneously assimilate and break down the organic matter. The process may be divided into two phases: acid and methane.

In the acid phase, facultative acid forming organisms convert the complex organic matter to organic acids (acetic, propionic, butyric, and other acids). In this phase little change occurs in the total amount of organic material in the system, although some lowering of pH results. The methane phase involves conversion of volatile organic acids to methane and carbon dioxide.

The anaerobic process is essentially controlled by the methane-forming bacteria. Methane formers are sensitive to pH, substrate composition, and temperature. If the pH drops below 6.0, methane formation essentially ceases, and more acid accumulates, thus bringing the digestion process to a standstill. Thus, pH and acid measurement constitute important operational parameters. The methane bacteria are highly active in the mesophilic ($27–43\,°C$) and thermophilic ($45–65\,°C$) ranges. Anaerobic digesters are most commonly operated in the mesophilic range. Recent thinking, however, is to operate the digester in the thermophilic range, for which the main advantages are increased efficiency and improved dewatering.

The anaerobic sludge digesters are of two types: *standard rate* and *high rate*. In the standard-rate digestion process the digester contents are usually unheated and unmixed. The digestion period may vary from 30 to 60 d. In the high-rate digestion process, the digester contents are heated and completely mixed. The required detention period is 10–20 d. Often a combination of standard- and high-rate digestion is achieved in two-stage digestion. The second stage digester is a standard-rate digester, that mainly

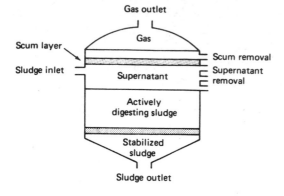

(a) Standard-rate sludge digestion

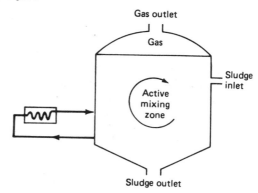

(b) High-rate sludge digestion

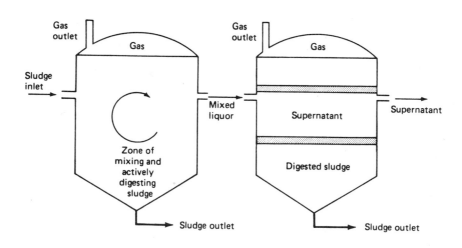

(c) Two-stage sludge digestion

Figure 9.3 Typical anaerobic sludge digesters. Source: adapted from Qasim (1994).

separates the digested solids from the supernatant liquor, although additional digestion and gas recovery may also be achieved. The schematic flow diagram of standard- and high-rate as well as single- and two-stage digestion processes are shown in Figure 9.3. The design criteria for anaerobic digester treating municipal wastewater sludge are provided in Table 9.7.

The most important factors controlling the design and operation of anaerobic digestion are digester capacity, digester heating and temperature control, mixing, gas production and utilization, digester cover, supernatant quality, and sludge characteristics (U.S. EPA 1979).

9.3.2 CHEMICAL AND PHYSICAL TREATMENT

The chemical and physical waste treatment processes utilize addition of chemicals into the wastewater to enhance removal of contaminants. It should be noted that chemical treatment is always used in conjunction with physical as well as biological treatment systems. As an example, activated sludge process utilizes oxygen transfer which is a chemical phenomenon.

TABLE 9.7. Typical Design Criteria for Standard-Rate and High Rate Digesters Digesting Municipal Wastewater Sludge.

Parameter	Standard Rate	High Rate
Solids retention time (SRT), days	30–60	10–20
Sludge loading, kg VS/m³ d	0.64–1.60	2.40–6.41
(lb VS/ft³d)	(0.04–0.10)	(0.15–0.40)
Volume criteria Primary sludge, m³/capita (ft³/capita)	0.03–0.04 (2–3)	0.02–0.03 (1–2)
Primary sludge + waste activated sludge, m³/capita (ft³/capita)	0.06–0.08 (4–6)	0.02–0.04 (1–3)
Primary sludge + trickling filter, m³/capita (ft³/capita)	0.06–0.14 (4–5)	0.02–0.04 (1–3)
Sludge feed solids concentration, percent dry wt.	2–4	4–6
Digested solids underflow concentration, percent dry wt.	4–6	4–6

Source: Adapted from U.S. EPA (1979) and Qasim (1994).

Likewise, coagulation and precipitation require flocculation and sedimentation which are physical unit operations. In considering the application of chemical unit processes for leachate treatment, it is important to note that these processes will be used in conjunction with either biological or physical treatment processes. Common chemical and physical treatment processes applied to leachate treatment are:

(1) Chemical treatment: (a) coagulation and precipitation, (b) Carbon adsorption, (c) Ion exchange, and (d) Chemical oxidation

(2) Physical treatment: (a) Natural evaporation, (b) Air stripping, (c) Flocculation and sedimentation, (d) Filtration, and (e) Reverse osmosis and ultrafiltration

Most of these processes are briefly described below. For detailed discussions readers should refer to textbooks that provide the theory and design of these processes (Berkowitz et al. 1978; Tchobanoglous and Schroeder 1985; Martin and Johnson 1987; Metcalf and Eddy, Inc. 1991; Qasim 1994).

9.3.2.1 Coagulation, Flocculation, Precipitation and Sedimentation

Coagulation, flocculation, precipitation and sedimentation are fully-developed processes currently used in a wide variety of industrial processes and waste treatment applications. While flocculation, precipitation and sedimentation are individual process steps, they are interrelated, and are often combined into a single overall treatment process. They can be readily applied to a variety of aqueous hazardous and industrial wastes for the removal of precipitable substances, such as soluble heavy metals, dissolved organics, and for the removal of colloidal particles suspended in the liquid.

Flocculation is the process whereby small, unsettleable colloidal particles in a liquid are made to agglomerate into larger more settleable particles. The process utilizes chemical dispersion in a rapid mix basin, followed by flocculation to agglomerate colloids into floc. Precipitation is a physio-chemical process where a substance in solution is transformed into a less soluble or insoluble phase. Sedimentation is a purely physical process whereby particles suspended in a liquid are made to settle by means of gravitational and inertial forces acting on both the particles suspended in the liquid, and on the liquid itself. These processes have long been widely used for a variety of municipal and industrial applications. Examples of such applications are turbidity removal, water softening, phosphorus and heavy metals removal.

Precipitation, flocculation and sedimentation are generally practical, effective and relatively low-cost processes for the removal of precipitable soluble substances and suspended particles from aqueous waste streams. These processes require chemical feeders, rapid mixing, flocculation and sedimentation basins, and are particularly suitable for the treatment of high volume/low concentration waste streams. In determining the applicability of precipitation, flocculation and sedimentation to a specific waste stream, simple laboratory treatability tests are usually needed to develop the design information, and to evaluate the performance.

9.3.2.2 Carbon Adsorption

A large variety of organic solutes, and a more limited number of inorganic solutes can be removed from aqueous waste streams by adsorption onto activated carbon. Activated carbon has a high adsorptive surface area (500–1500 m^2/g). It is used as powdered activated carbon (PAC) or as a granular activated carbon (GAC) bed. There are many full-scale carbon adsorption systems currently in use for industrial/municipal wastewater treatment. In general, the process works best with chemicals that have a low water solubility, high molecular weight, low polarity, and low degree of ionization. Some nonbiodegradable organics are also removed by activated carbon. For aqueous waste streams containing refractory or toxic organics, carbon adsorption is an excellent, proven process. For more concentrated waste streams, solute removal would be even more efficient, but the more frequent regeneration required would significantly increase the costs. Development of new methods of regeneration, particularly ones that allow recovery of the adsorbate, could greatly expand potential industrial applications of the process.

Powdered activated carbon in conjuction with suspended growth biological reactor (under the name PACT) has been recently introduced for treatment of wastes containing biodegradable and toxic organics.

9.3.2.3 Ion Exchange

Ion exchange is well suited for general and selective removal of heavy metals and toxic anions from dilute aqueous waste streams. The process involves the interchange of ions between an aqueous solution and a solid material (the "ion exchanger" or "resin bed"). After removal of undesirable ions from the solution and exhaustion of the bed, the regeneration cycle is achieved by exposure to a second aqueous solution of different composition which removes the ions picked up by the exchanger. The process is most frequently carried out by pumping the waste stream through one or more fixed beds of exchanger.

Full-scale operations include cleanup of dilute solutions from electroplating and other metal-finishing operations, recovery of effluents from fertilizer manufacturing, and industrial deionization. Promising applications include removal of cyanides from mixed streams, and use of newer exchangers for selective removal of heavy metals. The dilute purified product stream is dischargeable to the environment. The regenerant stream requires further treatment for recovery or disposal.

9.3.2.4 Chemical Oxidation

Chemical oxidation is used for the destruction of cyanides, phenols and other organics, and precipitation of some metals. The treatment technology for its large-scale industrial applications are well established. The oxidation-reduction, or Redox reactions are those in which the oxidation state of at least one reactant is raised while that of another is lowered. Chemical oxidation should be considered for dilute aqueous streams containing hazardous substances, or for removing residual traces of contaminants after treatment. Chemical oxidation should be considered as a first treatment step when it contains constituents not amenable to other treatment methods, or as a final step to remove traces of contaminants after other treatment.

The process train for the chemical oxidation process includes adjustment of pH of the solution. The oxidizing agent is added gradually and mixed thoroughly. The oxidizing agent may be in the form of a gas (e.g., ozone, chlorine), a liquid (e.g., hydrogen peroxide), or a solid (e.g., potassium permanganate). Because some heat is liberated, concentrated solutions may require cooling, and may require careful monitoring and handling to avoid violent reactions. Application to industrial wastes is well developed for oxidation of organics and inorganics in dilute waste streams. In addition to the already established applications for removal of hazardous substances, chemical oxidation may be used to remove chlorinated hydrocarbons and pesticides from dilute streams. Laboratory and pilot studies have demonstrated the potential for chemical oxidation for treatment of hazardous waste streams.

9.3.2.5 Natural Evaporation

Natural evaporation or evaporation ponds are used to concentrate wastes and sludges. The process involves impoundments with no discharge. The amount of evaporation from the water surface depends on temperature, wind velocity, and humidity. There are substantial variations in the average daily evaporation rate from month to month and year to year. However, depending on the climatic conditions, large impoundment may be necessary if precipitation exceeds evaporation over several months. Therefore, considerations must be given to net evaporation, storage requirements, and possible percolation and groundwater pollution.

Natural evaporation has been used for the treatment of industrial wastewater in arid climates, particularly for disposal of brines. This method is particularly beneficial where recovery of residues is desirable. Impervious liners are necessary in most instances.

9.3.2.6 Air Stripping

Air stripping is used to remove volatile compounds such as ammonia and VOCs. The process requires pH adjustment followed by application over the stripping tower. Air is blown from the bottom. Air-water contact and droplet agitation by countercurrent circulation of large quantities of air through the tower provides opportunity for VOCs to escape from the liquid into the air stream. Stripping towers are simple to operate and can be very effective in removal of VOCs, but their efficiency is dependent on air temperature. There is however, concern over release of VOCs into the atmosphere. Several stripping methods would permit recovery of the ammonia from more concentrated solutions.

Removal efficiencies of VOCs over 90% have been obtained. The major operational disadvantage of stripping is calcium carbonate scaling of the tower.

9.3.2.7 Filtration

Filtration is a well developed liquid/solid separation process currently applied to polish the secondary effluent, or to produce an effluent that is low in suspended solids. Filtration is used as a pretreatment device prior to other advanced treatment processes.

Filtration consists of passing wastewater through a filtering medium that can strain out the colloidal particles. The filtering media may be fine sand, anthracite, mixed media, multi media, diatomaceous earth, or filter fabric. The filtration system may be operated under gravity or pressure filter. The solids accumulate in the media and therefore backwashing using clean water in the opposite direction of flow must be accomplished periodically to control the head loss through the filtration system.

Microstrainers are commonly used for polishing the effluent from biological or physical-chemical treatment processes. Microstrainers consist of a woven stainless steel or polyester plastic fabric having openings ranging from 15 to 60 μm, mounted over a rotating drum that is held horizontally in a tank. The influent enters the drum interior through one end and flows out through the filtering fabric. The solids are retained in the inside of the rotating screen and are washed by a jet of filtered water on the top of the entire length of the drum. The washwater is collected in a hopper and returned to the plant. The washwater is 3–5 percent of the total effluent volume. Major equipment for filtration devices include filter housing, filter media, underdrain system, and backwash system. The basic design factors for the filtration device are average design flow, filter media, filtration rate, applied head, allowable head losses, and backwash flow rate.

9.3.2.8 Reverse Osmosis and Ultrafiltration

Reverse osmosis is a demineralization process applicable to the production of high-quality water. The process consists of permeating liquid through a semipermeable membrane at pressures up to 10,000 kN/m² (1500 lb/in²). The membranes reject most of the ions and molecules while permitting acceptable rates of water passage.

The heart of the reverse osmosis process is a semipermeable membrane. Many types of membranes have been developed, but cellulose acetate and polyamide (nylon) are currently the most widely used membrane materials. Reverse osmosis modules suitable for water treatment involve the arrangement of membranes and their supporting structures so that the feed water under high pressure can pass through the membrane surface while product water is collected from the opposite face without brine contamination. Four different types of module designs have been developed: plate and frame, large tubes, spiral wound, and hollow fine fiber. High-quality feed is required for the efficient operation of a reverse osmosis unit. Pretreatment of a secondary effluent with filtration and carbon adsorption is usually necessary. pH adjustment (4.0–7.5), and hardness, iron and manganese removal is necessary to decrease scale formation. The brine is returned to the plant or handled separately.

Ultrafiltration is a process similar to reverse osmosis. It removes dissolved particles of larger size (0.002–10 μm range). Ultrafiltration is actually a physical screening process using a relatively coarse membrane. The applied pressure is normally below 1000 kN/m² (150 lb/in²).

The major equipment for a reverse osmosis system includes a membrane module with support system, pretreatment system, high-pressure pump and piping, and brine-handling and disposal system. The design parameters for reverse osmosis systems are pretreatment processes, average design flow, salinity, applied pressure, product recovery rate, rejection rate, and influent quality to the membrane.

9.4 TREATMENT PROCESSES APPLIED TO LEACHATE TREATMENT

Numerous treatability studies have been conducted with landfill leachates over the past three decades. Early studies utilized bench scale and pilot plant setup. Many treatment processes were tested and operational ranges and performance levels were established. In this section the performance levels of many biological, physical, and chemical treatment processes as applied to landfill leachate treatment are presented.

9.4.1 BIOLOGICAL TREATMENT PROCESSES

Many researchers have investigated the performance of aerobic and anaerobic treatment processes for leachate treatment. The objective was to evaluate the performance of the processes in terms of organics and heavy metals removal, and operational features such as foaming, metal toxicity, nutrient deficiency, and sludge settling. These findings are summarized below.

9.4.1.1 Activated Sludge

Activated sludge process successfully treated the leachate in most cases. BOD and COD removal efficiency varied from 90 to 99 percent. Lu et al. (1984) analyzed the laboratory data of many researchers. The following generalizations can be made in treating leachate by activated sludge process (Boyle and Ham 1974; Cook and Foree 1974; Chian and DeWalle 1976; Palit and Qasim 1977; Uloth and Mavinic 1977; Memoin et al. 1981; Zapf-Gilje and Mavinic 1981; Mavinic 1984; McClintock et al. 1990; Christensen et al. 1992):

(1) BOD and COD removal, 90–99 percent

(2) Metal removal 80–99 percent

(3) Operational range: (a) MLVSS concentration of 5000–10,000 mg/L, (b) Food to microorganisms ratio 0.02–0.06 per day, (c) Hydraulic retention time 1–10 days, (d) Solids retention time 15–60 days, and (e) Nutrient requirements (BOD_5:N:P::100:3.2:0.5)

The operational parameters clearly indicate that large amounts of organic matter in leachate were not readily oxidized, and required extensive biological activity for stabilization. Many problems were also reported. Among these are: (1) excessive aeration in conjunction with high metal concentrations contributed to foaming; (2) antifoaming agents and mechanical mixing reduced foaming; (3) biological activity was also affected due to nutrient deficiencies; (4) metals and other constituents in leachate inhibited biological activity; as a result, increased time was necessary for biostabilization.

Biological kinetic coefficients are used in the biological growth and substrate utilization rate equations, and are well accepted for developing the reactor design. Uloth and Mavinic (1977) and Mavinic (1984), compared the biological kinetic coefficients of aerobic treatment of leachate from many studies. The biological kinetic coefficients obtained from several studies are summarized in Table 9.8. A comparison of biological kinetic coefficients from various sources show remarkable consistency considering the highly variable and complex nature of the leachate.

Many researchers have attempted to develop the removal rates of various groups of organic compounds in an aerobic reactor. Chian and DeWalle (1977) studied aerobic biodegradation of leachate collected from a lysimeter having an age of 5 months. These researchers reported four distinct and successive phases of substrate utilization by microorganisms: (1) carbohydrates; (2) fatty acids; (3) catabolites such as amino acids; and (4) humic carbohydrate-like materials having a molecular weight greater than 50,000. The residual refractory organics remaining beyond the fourth phase consisted mainly of fulvic acid-like materials with a molecular weight predominantly in the range of 500–10,000. Based on these findings, these researchers suggested that leachates from young landfills in which organic matter contained high volatile acids can be readily treated by biological processes. Leachate from older landfills containing large proportions of refractory material would be more efficiently treated by chemical-physical treatment process (Chian and DeWalle 1976; DeWalle and Chian 1974a, 1974b). Fillos et al. (1990) conducted treatability studies on leachate generated from Fresh Kills landfill in New York. The leachate had low BOD to COD ratio (less than 0.1). This ratio is characteristic of old landfills containing mainly refractory organic compounds. Biological treatment in a sequencing batch reactor showed low BOD or COD removal. The ammonia concentration dropped from 750 mg/L to less than 1 mg/L. The alkalinity reduced at a rate of 6.3 mg/L as $CaCO_3$ per mg/L of ammonia nitrified. Dzombak et al. (1990) treated leachate in an extended aeration system. The BOD/COD ratio of the leachate was below 0.1. Different mean cell residence times (15–60d) were investigated. They observed that maximum COD removal of 40% was achievable at mean cell residence time of 60 days. McClintock et al. (1990) conducted treatability studies on leachates from several landfills. These investigators utilized conventional activated sludge, intermittent aerobic/anoxic (A/A) process, sequencing batch reactor (SBR), and modified Ludzack-Ettinger (MLE). The MLE system uses continuous A/A process.

The A/A process removed up to 70% of the nitrogen contained in leachate, compared to 20% nitrogen removal by a conventional activated sludge system, when the leachate BOD_5:TKN ratios exceeded 10. It was however, necessary to add an organic carbon source such as methanol during the anoxic cycles to achieve efficient denitrification when the BOD_5:TKN ratio was less than 10. The A/A system achieved greater nitrogen removals than the aerobic system with no loss in BOD removal efficiency, while using 50% less aeration energy than the fully aerobic system. The A/A system also removed iron to less than 1 mg/L at greater

TABLE 9.8. Summary of Kinetic Parameters Describing Carbon Removal from Landfill Leachates (Soluble Basis Only).

S_0 (mg/L)	k (d^{-1})	K_s (mg/L)	Y (mg/mg)	k_d (d^{-1})	θ_c (d)	Ambient temperature (°C)	Source
36,000–BOD$_5$	0.75	200	0.33	0.0025	6.5	23–25	Uloth and Mavinic (1977)
15,800–COD	0.60	175	0.40	0.05	—	22–24	Cook and Foree (1974)
13,640–BOD$_5$	0.77	20.4	0.39	0.022	3.6	23–25	Zapf-Gilje and Mavinic (1981)
	0.71	29.5	0.63	0.075	—	16	
	0.46	14.6	0.50	0.028	—	9	Graham and Mavinic (1979)
	0.29	11.8	0.43	0.008	7.5	5	
8,090–BOD$_5$	1.16	81.8	0.49	0.009	1.8	22–23	Wong and Mavinic (1984)
	1.12	63.8	0.51	0.018	1.8	15	
	0.51	34.6	0.51	0.006	4.0	10	
	0.34	34.0	0.55	0.002	5.4	5	
1,000–BOD$_5$	4.5	99	0.59	0.040	0.42	22–23	Lee (1979)
365–BOD$_5$	1.8	182	0.59	0.115	—	21–25	Palit and Qasim (1977)
3000–BOD$_5$	—	—	0.44	—	1–20	10	Robinson and Maris (1983)
2000–BOD$_5$	0.46	180	0.50	0.100	2–10	25	Gaudy et al. (1986)

Source: Adapted from Mavinic (1984); Lu et al. (1984); Robinson and Maris (1983); Gaudy et al. (1986).

247

than 99% removal efficiency. Manganese removal was greater than 50% with effluent concentrations of 4–5 mg/L. The metals were removed by oxidation to insoluble forms and incorporation of the insoluble precipitates into the bacterial floc. There was no difference in iron removal between the A/A system and a completely aerobic system, but the aerobic system achieved much more complete manganese removal. Manganese concentrations up to 50 mg/L were not inhibitory to the A/A system, and nickel concentrations up to 20 mg/L apparently were not inhibitory as well. Copper concentrations of 5 and 15 mg/L added to the old landfill leachate severely inhibited nitrification and denitrification and caused destruction of the solids in the MLE system. System recovery after inhibition by toxic metals was extremely slow, taking over one month. These investigators suggested that a pretreatment step should be included for leachates containing high iron and manganese concentrations (80 mg/L iron, 10 mg/L manganese) to reduce these concentrations, and to prevent low MLVSS/MLSS ratios and mixing problems in the activated sludge process.

The aerobic/anoxic/aerobic (A/A/A) sequencing batch reactor (SBR) process removed 72%–79% of the COD and TOC, 98% of the BOD_5 and 72% of the total nitrogen present in the leachate from an operating landfill. Over 98% of the iron was removed, leaving residual iron concentrations less than 1 mg/L. Up to 82% of the manganese was also removed, with residual concentrations of 1–2 mg/L. Zinc was removed almost down to detection limits (0.02 mg/L).

The MLE system with methanol addition removed up to 80% of the total nitrogen contained in the leachate from an old, out-of-service landfill. Over 94% of the iron was removed with residual iron concentrations less than 1 mg/L. Manganese in the leachate was reduced by about 34% with residual concentrations of about 0.6 mg/L. Nitrogen removal was increased by increasing the return flow.

Metal removal from leachate and its effect on the biological system has been of some concern. Uloth and Mavinic (1977), reported aluminum, cadmium, chromium, iron and manganese removal of 95 percent; and lead, nickel, and iron removals were within 76–85 percent. McClintock et al. (1990) also reported that MLVSS/MLSS ratio will decrease in a suspended growth reactor because of heavy metal precipitation. In their study, this ratio had decreased to less than 0.3. These investigators suggested that pretreatment may be necessary to reduce metal concentration prior to a biological waste treatment process.

9.4.1.2 Stabilization Ponds and Aerated Lagoons

Stabilization ponds and aerated lagoons have been effectively used for treatment and polishing of municipal wastewater. For pretreatment of leachate, they offer a relatively economical method, prior to disposal into municipal sewers or recycling into landfills.

Early studies were conducted to evaluate the performance of aerobic and anaerobic ponds, and aerated lagoons for treatment of raw leachate, and also to evaluate their performance in conjunction with other treatment processes. Chian and DeWalle (1977) summarized the comparative findings of many researchers in treating raw leachate and overflow from many other treatment processes by aerated lagoon. These comparative findings are provided in Table 9.9. These studies reported a COD removal of 22–99%. The range of BOD/COD, and TOD/TOC ratios fed to the aerated lagoon varied from 0.03 to 0.80, and 1.56 to 3.45 respectively. It is clearly noted that COD removal decreases for lower ratios. Chian and DeWalle (1977), treated high strength leachate in laboratory scale aerated lagoons. Nutrient addition was necessary. Removal efficiencies achieved were: 97% organics, over 99% iron, zinc, and calcium, and 76% magnesium. Vydra and Grimm (1976) studied the combination of aerated lagoon and polishing pond for leachate treatment. The aerated lagoon and polishing pond each had a 10-day aeration period. The combination provided an effective leachate treatment system.

9.4.1.3 Fixed Film Reactor

Lugowski et al. (1989) investigated the comparative performance of activated sludge and rotating biological contactor (RBC) in treating leachate from existing landfills. Bench scale and pilot plant studies indicated that BOD_5 removal rates in RBC were 95–97%, and soluble COD removal rates were 80–90%. The aeration basins showed significantly lower removal rates. The RBC unit after an aeration basin provided additional removal, and nitrification was 98–99%. A combination of biological (aeration and RBC) followed by chemical-physical treatment was necessary to achieve secondary effluent criteria. Fillos et al. (1990) reported that RBC effectively nitrified ammonia, but BOD or COD removal was quite low. In their study, the leachate from Fresh Kills landfill was treated which had BOD to COD ratio of less than 0.1. Dzombak et al. (1990) also treated leachate in a bench-scale RBC at hydraulic loadings of 5–15 m^3/m^2d (0.2–0.6 gpd/ft²). The BOD/COD ratio of leachate was less than 0.1. The RBC unit achieved essentially complete nitrification, but maximum COD removal was 40%.

Sanford et al. (1990) investigated the use of subsurface flow rock-reed filters for treatment of leachate from a solid waste landfill. Experiments under controlled environmental conditions using small-scale rock-reed filters (2.5 m long, 30 cm wide, and 30 cm deep) with 1 cm gravel showed excellent result. Common reed *phragmites sustralis* was used. The

TABLE 9.9. Performance of Aerated Lagoons for Leachate Treatment.

Initial COD, mg/L	BOD/COD	TOD/TOC	Leachate Treated	% COD removal	Detention time, d	Source
8,800	0.80	—	Raw	74	5 d	Boyle and Ham (1974)
15,800	0.05	3.45	Raw	98	10 d	Cook and Foree (1974)
3,550	0.64	3.20	Raw	77	0.6 d	Karr (1972)
500	0.52	1.56	Raw	58	0.3 d	Pohland (1972)
30,000	0.65	2.9	Raw	99	7 d	Chian and DeWalle (1977)
—	0.18	—	Anaerobic digester effluent	40	5 d	Boyle and Ham (1974)
510	—	2.53	Anaerobic filter effluent	22	1 d	Foree and Reid (1973)
1,000	—	2.35	Anaerobic filter effluent	17	7 d	Chian and DeWalle (1977)

Source: Adapted from Chian and DeWalle (1977).

leachate was medium strength with variable chemical quality, but highly biodegradable organics. The results showed BOD removal 76–94% with the greatest removal occurring in gravel minibeds. Removals of heavy metals, nitrogen and phosphorus were also high.

9.4.1.4 Anaerobic Treatment Processes

Anaerobic processes have also been used for treatment of landfill leachates. The process offers several benefits over aerobic processes. Among these are (1) sludge production is significantly reduced, (2) stabilization of organics is lower, and (3) recovery of methane may provide energy. Most researchers demonstrated BOD_5 removal in the range of 90–99% (Boyle and Ham 1974; Pohland 1975; Chian and DeWalle 1977). In these studies the average BOD/COD and TOD/TOC ratios were 0.68 and 2.86 respectively. Chian and DeWalle (1977) summarized the results of these studies (Table 9.10). Foree and Reid (1973) determined COD/TOC and BOD/COD ratios of effluent from anaerobic units treating leachate, and compared these with ratios of leachates from landfills having different ages. These results indicated that the effluent from an anaerobic treatment process is comparable to the leachate from a landfill of intermediate age. This clearly indicates that a substantial part of the biodegradable organic matter is already removed in the landfill. Therefore, biological treatment methods, and in particular, anaerobic processes are moderately effective in removing the remaining organic matter in the leachate.

Pfeffer (1986) provided a review of leachate treatment by anaerobic biofilm reactor. Due to high solids retention time (SRT), an anaerobic biofilm reactor is responsible for producing much better treatment efficiencies than a completely mixed anaerobic reactor. The metal precipitation on packing media and sludge accumulation in the void of the reactor may pose severe operational problems. Chian and DeWalle (1977) also reported plugging of the reactor. This problem is eliminated in the fluidized bed reactor (Pfeffer 1986). In a study, McClintock et al. (1990) attempted to treat leachate from an old landfill that had high refractory organics. Several processes were tried. The anaerobic filter was ineffective for reduction of COD.

Schafer et al. (1986) treated high strength leachate (BOD 38,500 mg/L and COD 60,000 mg/L) from an existing landfill using upflow fixed film reactor, and found 95% BOD and TSS removal. The removal of copper, lead and zinc were 88, 84, and 83% respectively. The organic loading in the reactor was 1.07 kg/m³ · d (67 lb BOD/1000 ft³ · d). The design details of the reactor are provided in Table 9.11.

TABLE 9.10. Performance of Anaerobic Digester for Leachate Treatment.

Initial COD, mg/L	BOD/COD	TOD/TOC	COD removal %	Detention time, d	Source
10,000	0.79	—	93	10	Boyle and Ham (1974)
12,900	0.45	2.81	92	10	Foree and Reid (1973)
16,500	0.62	2.92	99	15	Karr (1972)
5,500	.78	2.82	93	10	Karr (1972)
1,300	.81	—	87	1.2	Rodgers (1973)
30,000	0.65	2.90	97	27	Chian and DeWalle (1976)

Source: Adapted from Chian and DeWalle (1977).

TABLE 9.11. Anaerobic Filter Design Basis.

Design Parameter	Value
Minimum Hydraulic Retention Time, d	4.9
Average Hydraulic Retention Time, d	7.4
Anaerobic Reactor	
Volume, m³ (gal)	2813 (740,200)
Diameter, m (ft)	18.3 (60)
Side depth, m (ft)	10.7 (35)
BOD Loading, kg/m³ · d	
(lb/1000 ft³ · d)[a]	7.1 (442)
Media	
Depth, m (ft)	6.1 (20)
Volume, m³ (ft³)	215 (56,500)
Surface area, m²/m³ (ft²/ft³)	114.8 (35)
Volumetric void ratio	0.95
Maximum gas pressure, cm water (inches)	30.5 (12)
Design Flow	
Average m³/d (gpd)	380 (100,000)
Peak m³/d (gpd)	570 (150,000)

[a]Loading based on media volume, average flow and 30,000 mg/L BOD.
Source: Adapted from Schafer et al. (1986).

9.4.1.5 Combined Leachate and Municipal Wastewater Treatment

Combined leachate treatment in an existing municipal wastewater treatment plant is a convenient method. The requirements are: availability of sewer system, plant capacity to assimilate the waste, process compatibility with leachate characteristics, and facility to handle increased sludge production.

Boyle and Ham (1974), Chian and DeWalle (1977), Pohland and Harper (1985), Henry (1985), Kelly (1987), and Zachopoulos et al. (1990) conducted studies with combined leachate treatment. Boyle and Ham (1974) found that high strength leachate with COD exceeding 10,000 mg/L can be treated at a level of 5 percent by volume without seriously impairing the treatment process or the effluent quality. Beyond 5 percent by volume, leachate addition resulted in substantial solids production, increased oxygen uptake rates, and poorer settling of biomass. These researchers also suggested that the presence of metals, ammonia, other toxins, and extremely high organic load in leachate may cause severe upsets in the biological reactors. Chian and DeWalle (1977) also found that leachate greater than 4 percent by volume reduced the treatment plant efficiency.

Henry (1985) suggested that, when possible, the addition of leachate into a municipal sewer should be a preferred disposal method. He reported that studies by others indicated that high strength leachate (24,000 mg/L COD) when combined at less than 2% by volume with normal municipal wastewaters has no significant effect upon the performance of municipal wastewater treatment plants; however, if the percentage is increased to 5%, plant performance could deteriorate. Henry (1985) also noted that sludge volume generated from treatment of leachate alone is twice that from municipal wastewater, (1 kg SS/kg BOD_5). Also, aeration of leachate precipitated Fe, Cu, and other metal ions. In laboratory studies, Raina and Mavinic (1985) successfully treated combinations of 20 and 40% leachate by volume with municipal wastewater in aerobic batch (fill and draw) reactors with sludge residence times (SRT) of 5, 10, and 20 days. High SRTs improved settleability. Low temperatures showed a decline in treatment efficiencies and effluent quality. Zachopoulos et al. (1990) studied co-disposal of leachate in POTW and concluded that leachate can be treated without any adverse effect upon plant performance.

Barchyn (1984) reported a pilot test study on leachate combined with municipal wastewater where leachate was introduced at 10 and 20% by volume. Results indicated poor operational performance at a 10% dilution. Robinson (1985) and Birkbeck (1984) have reported calcareous precipitate formation in aerated biological treatment. The precipitates severely affected the operation of aerators and pumps. The precipitates coated pump and aerator impellers so severely that the motors were burned out, and monthly maintenance requirements included chipping off the tough white coatings with hammers and chisels.

Pohland and Harper (1985) reported greater success with combined leachate treatment in wastewater treatment plants. In their investigation BOD and COD removal efficiency (over 90 percent) and complete nitrification (over 80%) was obtained with 10-day sludge residence time. Chian et al. (1985) observed fairly efficient removals of heavy metals in aerobic biological treatment processes. Zinc, iron, cadmium, and manganese were easily removed, followed by lower removals of cadmium, lead and nickel. Kelly (1987) provided a complete literature review on the leachate treatment in municipal wastewater treatment plants. A summary of combined treatment of leachate in municipal wastewater treatment plants is provided in Table 9.12.

The literature on the subject of combined leachate and municipal wastewater treatment has many uncertainties about treatment plant performance. Although, BOD_5, COD and heavy metal reduction is well established, relative proportions of leachate effectively treated, effects of heavy metals, ammonia conversions, temperature effects, sludge production, foaming, poor solids settleability, heavy metal accumulation, and

TABLE 9.12. Summary of Combined Activated Sludge Leachate Treatment Results.

Treatment	HRT, d	SRT, d	F:M	MLSS, mg/L	TKN, mg/L[a]	TKN, %[b]	COD, mg/L[a]	COD, %[b]	BOD, mg/L[a]	BOD, %[b]	Comments	Reference
Activated sludge 95%	1-5	—	0.3-1.5[c]	—			—	—	1550-8000	2-93	Foaming, power needs, bulking sludge	Boyle and Ham (1974)
Activated sludge 95% wastewater, 5% leachate (by volume)	1	—	<0.27[c]	2500						High	Process appears workable as long as leachate is less than 10% by volume	
Activated sludge	10	—	0.14[c]	7000					9500	90-99	Foaming solved with defoamer; good setting sludge, no nutrients needed	Morgan (1973)
Activated sludge	6-10	—	—	—	10-100 effluent	75-90	500 effluent	95	100 effluent	95		Chian et al. (1985)
Combined wastewater >55% wastewater												Pohland and Harper (1985)

(continued)

255

TABLE 9.12. (continued).

Treatment	HRT, d	SRT, d	F:M	MLSS, mg/L	TKN, mg/L[a]	TKN, %[b]	COD, mg/L[a]	COD, %[b]	BOD, mg/L[a]	BOD, %[b]	Comments	Reference
Activated sludge	—	—	0.05–0.1[d]		(Org.N) 1000	—	>3000	—	>1000	>90	Filamentous growth at high loading rates	Ehrig (1985)
Activated sludge	—	—	0.05–0.1[d]		(Org.N) 300	—	<3000	—	<300	>90	3.57 mgCaCO$_3$ destroyed/mg org-N	Raina and Mavinic (1985)
Bench test once daily feed, fill & draw activated sludge, 80% wastewater, 20% leachate	—	5	0.63[d]	282			888	90.7	—	—	SVI decreased as SRT increased	
	—	10	0.32	513	—		812	91.1	—	—		
	—	20	0.16	987			781	90.0	—	—		
Bench test once daily feed, fill & draw activated sludge, 60% wastewater, 40% leachate	—	5	0.25	1029			1297	93.3	—	—	SVI was least at SRT of 10 days	
	—	10	0.17	1402			1160	94.3	—	—		
	—	20	0.14	1899			1304	93.9	—	—		

TABLE 9.12. (continued).

Treatment	HRT, d	SRT, d	F:M	MLSS, mg/L	TKN, mg/L^a	TKN, %^{b}	COD, mg/L^a	COD, %^{b}	BOD, mg/L^a	BOD, %^{b}	Comments	Reference
Activated sludge, 90% wastewater 10% leachate	0.75	20–30	—	<2000	—	—	110 (TOC)	36 (TOC)	—	— 75 (TOC)	MLSS were granular and settled well but fines remained in suspension	Barchyn (1984)
Activated sludge 80% wastewater 20% leachate	0.75	20–30	—	<2000	113	0–84	200 (TOC)		—	—		
Activated sludge (aeration basin)	—	20.6	0.28–0.32^e		348	98.0	13296	98.3	10721	99.7	Full strength leachate	Maris (1985)
	—	14.0	0.28–0.32^e		353	88.4	11929	96.2	9674	99.8	Metal content of dried sludge was high	
	—	10.7	0.28–0.32^e		590	78.3	17717	97.0	11713	99.7		
	—	7.0	0.28–0.32^e		374	96.5	11831	96.3	7650	99.5		
	—	5.0	0.28–0.32^e		388	54.9	6240	91.0	2944	98.7		
	—	3.0	0.28–032^e		87	96.5	4141	91.2	1663	99.3		

^{a} = Influent.
^{b} = Removed.
^{c}F:M - - - kgBOD/kgMLSS · d.
^{d}F:M - - - kgCOD/kgMLSS · d.
^{e}F:M - - - kgCOD/kgMLVSS · d.
Source: Adapted in part from Kelly (1987).

precipitates formation have all been observed to varying degrees. Many authors believe that leachate quality has some impact upon the performance of the treatment plant. Therefore, the performance must be investigated on a case by case basis. When toxins are expected, or where a high proportion of leachate is to be treated, a pretreatment facility may be considered. A simple aerated lagoon followed by polishing lagoon may serve the pretreatment needs.

9.4.1.6 Land Treatment

Land treatment of leachate was investigated by Nordstedt (1975) in Florida. Leachate was collected in a detention pond and was applied over pasture grasses through the sprinkler irrigation system. Examination of soil samples from various depths did not show any change in the quality of soil. Careful consideration must be given to soil properties, hydrogeologic conditions, and regulatory requirements for land treatment of leachate.

9.4.2 PHYSICAL AND CHEMICAL PROCESSES

Numerous studies have been conducted over 20 years to evaluate the performance of many physical and chemical treatment processes for treatment of both raw leachate and biologically treated leachate. Physical and chemical treatment processes are particularly useful in treating leachate from older landfills that have lower biodegradable organic carbon, or as a polishing step for biologically treated leachate. Chian and DeWalle (1977), and Keenan et al. (1983, 1984) conducted an extensive literature search on this subject and summarized the results of many investigators. These summary results along with the results of more recent studies are provided in Table 9.13. A brief description of many physical and chemical treatment processes as applied to leachate treatment is provided below.

9.4.2.1 Precipitation and Coagulation

Chian and DeWalle (1977) showed that lime precipitation predominantly removed the organic matter having the molecular weight larger than 50,000. This fraction as indicated earlier is present in relatively low concentrations in leachate from young fills, and almost absent in leachate from old fills. Since the proportion of organics of 50,000 molecular weight fraction was found to increase and then decrease in older fills, lime treatment was the most effective method for treatment of leachate from medium-age fills. This mechanism was further supported by the work of Ho et al. (1974), and Thornton and Blanc (1973). These researchers and others have further demonstrated that BOD and COD removal by lime precipitation is small, but removal of color, iron and other multivalent cations was excellent at higher lime doses (300–1000 mg/L).

Coagulation with alum and ferric chloride was also not very effective in removing BOD and COD. Removal of color, iron, and other multivalent cations was marginal. High chemical doses were needed and the process was sensitive to pH. Also production of large quantities of sludge was reported.

9.4.2.2 Ion Exchange

Pohland (1975), utilized combination of cation and anion exchange resins to improve the quality of biologically treated leachate. The process was effective in removing dissolved salts, and nutrients. Very little removal of residual organics was reported. Mixed resin bed was considered a promising approach for removal of the nonorganic fraction of leachate.

9.4.2.3 Carbon Adsorption

Removal of residual organics from leachate by activated carbon was investigated by many researchers. Findings of many researchers are provided in Table 9.13.

Activated carbon treatment of raw leachate generally gave better removal of organic matter than is observed with chemical precipitation. Using relatively large dosages of powdered activated carbon, COD removals between 34% and 85% were obtained. Column studies with activated carbon resulted in COD removals of 59%–94%. Chian and DeWalle (1977), reported that leachate having a COD/TOC ratio of 2.9, showed an initial TOC removal of 70%, which decreased to 13% after 140 bed volumes. This clearly indicated that treatment of leachate by activated carbon is not feasible due to the large quantities of activated carbon required.

Chian and DeWalle (1977) suggested that the preceding studies were conducted using leachate obtained mostly from young fills. Since free volatile fatty acids are the main components in such leachate, the overall removal of organics by activated carbon will be lower than the more effective biological treatment processes discussed earlier. Kipling (1965) reported that activated carbon removed acetic, propionic, and butyric acids by 24%, 33%, and 60%, respectively, at a dosage of 5,000 mg/L. Burchinal (1970) suggested that variation in COD or TOC removal by activated carbon in young leachate is due to different magnitudes and proportions of low and high molecular weight free volatile fatty acid fractions in the leachate.

Cook and Foree (1974) found that activated carbon treatment provided higher COD removals from biologically pretreated leachate than that from the raw leachate (Table 9.13). Chian and DeWalle (1977) however, reported up to 70 percent COD removal by activated carbon with stabilized leachate

TABLE 9.13. Results of Treatment Efficiencies Obtained in Different Physical-Chemical Treatment Studies.

Treatment Process	Initial COD mg/L	BOD/COD	COD/TOC	Treatment System	COD Removal %	Dosage	Reference
Chemical Precipitation	14,900	0.45	3.45	Lime	13	2,760 mg/L Ca(OH)$_2$	Cook and Foree (1974)
	9,100	0.75	—	Ferric chloride	16	1,000 mg/L	Ho et al. (1974)
	9,100	0.75	—	Alum	5	1,000 mg/L	
	10,800	0.74	—	Lime	4	1,840 mg/L	
	558	0.27	—	Lime treatment of anaerobic digester effluent	8	2,700 mg/L	
	366	0.11	—	Lime treatment of anaerobic digester effluent polished by aereated lagoon	29	1,400 mg/L	
	4,800	0.66	2.73	Alum and lime	40	2,250 mg/L Al$_2$(SO$_4$)$_3$ and 800 mg/L CaO	Karr (1972)
				Ferrosulfate	13	2,500 mg/L FeSO$_4$ 7H$_2$O	
	3,400	0.81	—	Lime	0	1,000 mg/L	Rogers (1973)
	1,240	0.66	2.78	Lime and aeration	8	210 ml saturated lime/L leachate	Simensen and Odegaard (1971)
	1,234	0.68	2.88	Iron and aeration	0	200 mg/L FeCl$_3$	
	1,234	0.68	2.88	Alum and aeration	11	180 mg/L Al$_2$(SO$_4$)$_3$	

TABLE 9.13. (continued).

Treatment Process	Initial COD mg/L	BOD/COD	COD/TOC	Treatment System	COD Removal %	Dosage	Reference
	5,033 12,923	0.60 0.57	— —	Lime Lime	24 26	1,350 mg/L 1,200 mg/L	Thornton and Blanc (1973)
	2,000	0.36	—	Alum	31	2,700 mg/L	Van Fleet et al. (1974)
	2,820	0.65	2.89	Lime	26	450 mg/L	Chian and DeWalle (1977)
	2,000	—	0.10	Alum and cationic polymer	45 25	200 mg/L alum and 330 mg/L polymer 200 mg/L alum only	Fillos et al. (1990)
Activated Carbon and Ion-Exchange Adsorption	330	0.07	2.57	Activated carbon batch treatment of aerated lagoon effluent	70	—	Cook and Foree (1974)
	3,290	0.45	3.45	Activated carbon column treatment of lime pretreated leachate	81	15 min HRT after initial volume	Cook and Foree (1974)
	4,920 7,213	0.75 0.75	— —	Activated carbon, batch Activated carbon, column	34 59	16,000 mg/L 45 min HRT after 3 volume turnovers	Ho et al (1974)
	5,500	0.66	2.73	Activated carbon, batch	60	160,000 mg/L	Karr (1972) Pohland and Kang (1975)

(continued)

TABLE 9.13. (continued).

Treatment Process	Initial COD mg/L	BOD/COD	COD/TOC	Treatment System	COD Removal %	Dosage	Reference
Activated Carbon and Ion-Exchange Adsorption (continued)	184	0.18	1.50	Carbon batch treatment of activated sludge effluent	91	10,000 mg/L	Van Fleet et al. (1974)
	120	0.18	1.50	Ion exchange treatment of activated sludge effluent	58	5,000 mg/L cation and anionic resin mixture	
	2,000	0.36	—	Activated carbon column treatment of leachate	71		
				Activated carbon column treatment of alum pretreated leachate	94		Chian and DeWalle (1977)
	632	0.65	2.89	Activated carbon column treatment of leachate	70		
	346	<0.1	2.55	Activated carbon column of effluent of aerated lagoon	70		
	527	<0.1	2.46	Ion exchange column treatment of effluent of aerated lagoon	50		
	932	—	2.90	Activated carbon column treatment of effluent of anaerobic filter	50		
	522	<0.1	2.70	Activated carbon column treatment of aerated effluent of anaerobic filter	70		
	2,000	<0.1	—	PAC, 119 mg/L TOC/g carbon	85		Fillos et al. (1990)

TABLE 9.13. (continued).

Treatment Process	Initial COD mg/L	BOD/COD	COD/TOC	Treatment System	COD Removal %	Dosage	Reference
Chemical Oxidation	330	0.07	2.57	Chlorination	33	65 ml bleach/ L sample	Cook and Foree (1974)
	1,500	0.75	—	Chlorination with calcium Ca(ClO)$_3$ hypochlorite	8	8,000 mg/L	Ho et al. (1974)
	7,162	0.75	—	Ozonation	37	4 hr, 7,700 mg O$_3$/L•hr	
	4,800	0.66	2.73	Chlorination	22	2,000 mg/L Cl$_2$	Karr (1972)
	139	0.04	2.10	Ozonation	22	4 hr, 34mg O$_3$/L•hr	Chian and DeWalle (1977)
	1,250	—	2.90	Ozonation of anaerobic filter effluent	37	3 hr, 600 mg O$_3$/L•hr	
	627	—	2.50	Ozonation of aerated lagoon effluent	48	3 hr, 400 mg O$_3$/L•hr	

(continued)

263

TABLE 9.13. (continued).

Treatment Process	Initial COD mg/L	BOD/COD	COD/TOC	Treatment System	COD Removal %	Dosage	Reference
Reverse Osmosis	53,300	0.65	2.89	Reverse osmosis of leachate at pH 5.5, cellulose acetate membrane	56	50% permeate yield	Chian and DeWalle (1977)
	53,300	0.65	2.89	Reverse osmosis of leachate at pH 8.0, cellulose acetate membrane	89	50% permeate yield	
	900	—	2.90	Reverse osmosis of anaerobic filter effluent	98	77% permeate yield	
	536	—	2.50	Reverse osmosis of aerated lagoon effluent, cellulose acetate membrane	95	50% permeate yield	

Source: Adapted in part from Chian and DeWalle (1976, 1977).

having a BOD/COD ratio of less than 0.1. In this study the reported carbon concentration was 0.17 mg COD/mg activated carbon. McClintock et al. (1990) reported GAC at a dose of 2 g/L reduced effluent COD concentration by 84% after 40 h of contact time. The effluent treated in this study was obtained from a modified Ludzack-Ettinger process (anoxic/anaerobic). PAC added directly into the process train at a rate of 1 to 3 g/L reduced COD concentration to 40 and 19 mg/L respectively. These investigators also reported that GAC dose of 1 mg/L reduced refractory COD concentrations of SBR effluent by 30% after 17 hours. GAC was also effective in reducing the turbidity of this effluent. Pohland (1975) studied carbon adsorption of leachate that was previously treated by ion exchange resins. The carbon treatment was moderately successful in removing COD and other constituents. It was concluded that if leachate is treated by both carbon and mixed resins, the carbon treatment should precede the mixed resins. Ho et al. (1974) demonstrated 55 percent COD removal, and complete color and odor removal by activated carbon column.

Pirbazari et al. (1989) investigated a combined adsorption and biodegradation process in the form of a biologically active granular activated carbon absorber. Good removal of organics was observed by the biofilm that formed over the carbon surface. A mathematical model provided good agreement.

9.4.2.4 Chemical Oxidation

Several investigators studied chemical oxidants including chlorine, calcium hypochlorite, potassium permanganate and ozone for treatment of leachate. Ho et al. (1974) showed reasonable success in removing COD, iron and color at high doses of oxidants. Studies by Chian and DeWalle (1977) suggested that calcium hypochlorite produced the best results. Ozone treatment of leachate from young landfill was not promising because of strong resistance of fatty acid to ozonation. However, almost 22 percent COD removal was obtained by ozonation of leachate from an old fill.

9.4.2.5 Reverse Osmosis

Of all the physical-chemical processes evaluated for leachate treatment, reverse osmosis was found most effective for removal of COD. Chian and DeWalle (1977) reported between 56–70% removal of TOC with the conventional cellulose acetate membrane, while the removal increased to 88% with the polyethyleamine membranes. Since low rejection of undisassociated fatty acids by membranes was responsible for the leakage of TOC into the permeate, the performance of the membrane process was improved to 94% when the pH of the leachate was increased from 5.5 to 8.0

(Chian and Fang 1973). In addition to efficient TOC removal, total dissolved solids removal was as high as 99%. In most studies, severe membrane fouling was experienced, and biological pretreatment of leachate prior to membrane processes was necessary. Lu et al. (1984) suggested that reverse osmosis is perhaps most effective as a post biological treatment step for removal of residual COD and dissolved solids. Chian and DeWalle (1977) also noted that membranes are sensitive to pH.

Kinman et al. (1985) studied reverse osmosis membrane fouling by sanitary landfill leachate and copper plating solutions. Silt Density Index (SDI) and permanganate demand tests were used to determine if the process could be operated for an extended period of time without fouling. The facility operated for the six month period of the project. The SDI test provided no useful information, but permanganate demand exhibited strong predictive abilities. The permeate from the leachate was clear but still possessed some odor. Flushing of membrane restored permeate flux.

Kinman et al. (1985) also reported that the reverse osmosis treatment of landfill leachate reduced total dissolved solids from 4000 mg/L to 100 mg/L with 97% reduction in organics. Metal concentrations were reduced below detection limit. A permeate flux of 2.50 m³/m² · d (6 gpd/ft²) was best maintained around at pressures 1400–1700 kN/m² (200–250 psi). There were no major problems encountered in operation of RO unit for treatment of leachate.

Slater et al. (1983) investigated different treatment trains in conjunction with RO to treat landfill leachates. Pretreatment consisted of the removal of bulk oil by gravity separation, coagulation by lime, recarbonation, and pH adjustment. Leachate feed to RO unit had 16,400 mg/L TDS, 26,400 mg/L COD and 8,500 mg/L TOC. The RO unit had a permeate flux of 180 L/m² · d (4.4 gpd/ft²). The rejection rate of TDS, COD, and TOC were 98, 68, and 59% respectively. These researchers also investigated lime coagulation, recarbonation, sedimentation, biological treatment and filtration prior to RO. The permeate flux increased to 290 L/m² · d(7.1 gpd/ft²). Rejection rates of TDS, COD, and TOC increased considerably.

Whittaker et al. (1985) investigated RO in a self contained mobile unit to demonstrate the technology. The unit was tested for municipal and hazardous landfill leachate treatment. High removals of organic contaminants such as dichloroethane, dimethyl ether and dioxin were obtained.

Krug and McDougal (1988), and Smith and Krug (1990) investigated a two-stage process for treatment of landfill leachate. The process involved (1) pretreatment by precipitation and microfiltration for removal of toxic metals and suspended solids, and (2) reverse osmosis to remove residual organics and inorganics. The process train included flow equalization, chemical addition, microfiltration, and reverse osmosis. The sludge from microfiltration was dewatered in plate and a frame filter press. The RO concentrate was recirculated. The system was tested for numerous chemical feed conditions. The microfiltration system achieved stable flux rates ranging from 92 to 132 L/min · h, and RO membranes achieved stable flux of 32 L/min · h. Overall process removal for key contaminants of concern were generally over 90% at waste volume reductions of 75–80% (Smith and Krug, 1990). The result demonstrated the feasibility of two-step process for treatment of leachate.

9.4.3 LEACHATE RECIRCULATION

One of the most innovative and acclaimed methods of leachate treatment is the circulation of high strength leachate back to the landfill, so that it may again percolate through the refuse. Recirculation essentially uses the landfill as a generally uncontrolled anaerobic digester for effective anaerobic treatment. By recirculating the leachate, the organic component of the leachate can be reduced by the biological communities active within the refuse mass. Using experimental landfill columns, Pohland (1972, 1975, 1980) and Pohland et al. (1990) found that when leachate was continuously collected and reapplied to the landfill surface, the organic load of the leachate decreased to a small fraction of its peak value (i.e., from 20,000 mg/L COD to less than 1,000 mg/L) in a period of just over a year. An investigation of the effects of leachate recycle using a pilot-scale fill produced similar results (Pohland, 1979). Leckie et al. (1975, 1979) also conducted field tests with leachate recirculation and reported enhanced leachate stabilization. The results of leachate recirculation studies by Pohland (1975), and Leckie et al. (1979) are shown in Figure 9.4. Both studies clearly indicate rapid decline in COD concentration. Both Pohland (1975) and Leckie et al. (1975 and 1979) suggested that leachate recirculation provides for a more rapid development of an active anaerobic bacterial population of methane forming bacteria. The rate of removal of organic components was further enhanced by the initial addition of sewage sludge and by pH control.

In addition to rapid decline in leachate COD concentrations over time provided by recirculation, similar concentration histories were observed for BOD, TOC, volatile acids, phosphate, ammonia-nitrogen, and TDS. Reduction in nitrogen and phosphorous were low, and recirculated leachate remained high in inorganic material. Leachate recirculation offers several advantages in terms of leachate control. Among these are:

- acceleration of landfill stabilization
- sustained reduction in leachate organic components
- possible volume reduction due to evaportranspiration
- delay in the starting time for leachate treatment
- reduction in leachate treatment costs

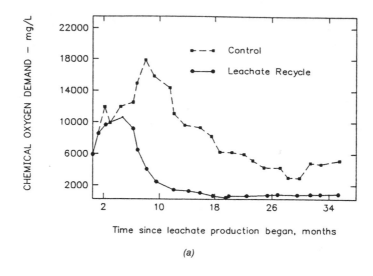

(a)

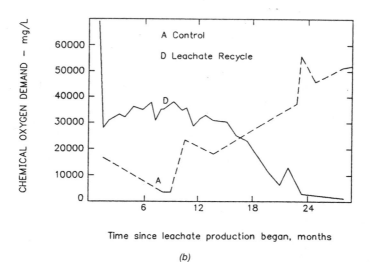

(b)

Figure 9.4 Effect of leachate recirculation upon COD reduction. (a) Pohland (1975), (b) Leckie et al. (1979).

Cited disadvantages are the high capital and maintenance costs of leachate recirculation systems, and odor problems with above-grade systems.

Tolman et al. (1978) provided design and operation information on three types of leachate recirculation systems. These are spray irrigation, overland flow, and subgrade irrigation. A summary of these systems is given in Table 9.14. Lu et al. (1984) provided a brief description of these systems. These descriptions are provided below:

Spray irrigation is accomplished by periodically pumping collected effluent through spray nozzles placed at 15 to 30 m (50 to 100 ft) intervals on the landfill. Spray irrigation has the advantage that some leachate treatment may be achieved during spraying as a result of both aeration and infiltration through the soil cover. Overland flow irrigation is an expensive well-developed technique by which trenches, spreading basins, or perforated pipes are used to distribute the effluent. Leachate is periodically pumped into the distribution system and allowed to infiltrate into the soil. With both spray irrigation and overland irrigation, an impermeably lined pond is necessary to provide storage of the leachate between irrigation periods. The overland flow irrigation as a potential treatment process represents a viable means of treating recycled leachate.

Subgrade irrigation is also considered a potential means to promote uniform stabilization of the fill. In these approaches, leachate is distributed throughout the fill through smaller subgrade tile fields or pipes. Very large tile fields must be subdivided into smaller units and each unit is fed with leachate separately. Subgrade irrigation may be used continuously and has the advantage of avoiding local odor problems.

9.4.4 Leachate Treatment System Case Histories

Leachate treatment requires integration of several treatment processes to achieve the desired level of treatment. Because leachates can have high strength initially, and both flow and strength change greatly with time, the process train must incorporate such changes. Two leachate treatment system options are: (a) complete treatment for effluent discharge into the environment, and (b) pretreatment followed by combined treatment at municipal wastewater treatment facility. Over the past 25 years, many treatment systems have been designed and constructed. In this section, a review of case histories of many wastewater treatment facilities for leachate treatment, and experience gained over years are presented. Readers should refer to the references cited for more information on each system described.

9.4.4.1 Tullytown, Pennsylvania

Steiner et al. (1979), and Keenan et al. (1984) presented a complete treatment system for a landfill in Tullytown, Pennsylvania. The leachate was collected from a fill that had an asphalt liner. The process train included flow equalization, lime precipitation and clarification, ammonia stripping, acti-

TABLE 9.14. Leachate Recirculation and Collection Technologies.

Control Techniques	Optimum Site Conditions	Limitations	Perceived Effectiveness
Spray irrigation	High evaporation rates for the site; site vegetation must be well established. Site contour minimizes ponding, runoff, etc.	Practical only in dry months; wet weather conditions minimizes effectiveness. Pumping equipment, collection wells and associated piping is required.	Most effective in arid regions, or during the summer months when treatment efficiency is highest.
Overland flow	Land availability with existing leachate collection, spray equipment, and ditches for recycle are needed. Existence of rooted grasses and slope conducive to natural gravity flow, favorable soil properties, and high evaporation/evapo-transpiration rates are desired.	Large area requirements; leachate collection and distribution system required. Conducive weather conditions required for optimum use.	Provides an effective treatment scheme when land is available and climate conditions are favorable.
Subgrade irrigation	Collection and leachate distribution systems for the fill are required.	Elaborate piping and distribution system required.	Provides an effective landfill stabilization mechanism. Largely unaffected by climatic conditions. Treatment efficiency is less than for spray irrigation or overland flow.

Source: Tolman et al. (1978), and Lu et al. (1984).

vated sludge and chlorination. The design is illustrated in Figure 9.5. A summary of the plant operating data appears in Table 9.15.

Because of low flow conditions of the receiving waters, treated leachate was discharged only from December until April. In other months, the treated leachate was recycled to the landfill. Major problems reported for this operation included leachate flow and strength variability. Leachate COD varied over a wide range from day to day, though concentrations averaged between 10,000 and 15,000 mg/L during the first three years of operation. A lime precipitation stage was needed to reduce the metals concentrations to the levels which would not inhibit the activated sludge process. Similarly, the toxic effect of ammonia was controlled by the ammonia stripping process. After lime treatment and ammonial stripping, sulfuric and phosphoric acids were added to lower the pH and raise nutrient levels before the activated sludge process. The activated sludge process reduced BOD and COD concentrations by over 90 percent. An aeration period of more than one day was provided. Steiner et al. (1979) reported that the system operation was sensitive, and effluent standards approached only during optimum operation of the plant. Although the system operated satisfactorily for high strength leachate, it was clear that a sophisticated treatment system was needed.

TABLE 9.15. Summary of Operational Data, Tullytown, Pennsylvania.

Parameter	Raw Leachate mg/L	Final Effluent, mg/L	Removal, Percent	Discharge Standard, mg/L
Ammonia-N	758	75	90.1	35
BOD$_5$	11,886	153	98.7	100
Cadmium	0.08	0.017	78.2	0.02
Chromium	0.26	0.07	73.1	0.1
COD	18,490	945	94.9	*
Copper	0.40	0.11	72.5	0.2
Iron	333	2.7	99.2	7.0
Lead	0.74	0.12	83.8	0.1
Mercury	0.006	0.004	27.4	0.01
Nickel	1.76	0.75	57.4	*
Zinc	19.5	0.53	97.3	0.6

*No discharge standard for this parameter.
Source: Adapted from Steiner et al. (1979).

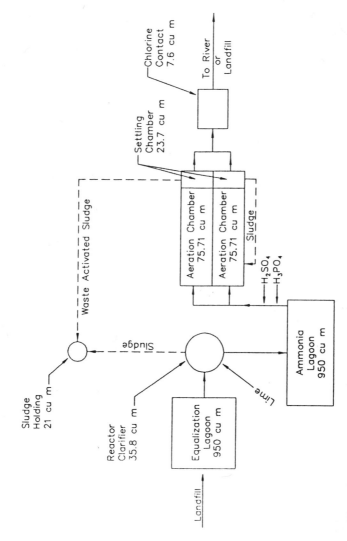

Figure 9.5 Schematic of complete leachate treatment at Tullytown, PA. Source: adapted from Steiner et al. (1979).

9.4.4.2 New Hampshire Facility

The Turnkey Recycling and Environmental Enterprises operates a landfill leachate treatment facility near Rochester for Waste Management of New Hampshire (Anonymous 1992). The facility utilizes *Zimpro Passavant 2-stage PACT system*. The process train utilizes flow equalization, and chemical addition followed by inclined plate separator. In this unit chemical precipitation removes metals from leachate. Next, the leachate passes to the two-stage PACT system. Powdered activated carbon is combined with biological treatment. The first stage PACT system is combined with the anaerobic biological treatment followed by PACT system combined with aerobic process. The facility is designed for a flow of 150 m³/d (40,000 gpd). The PACT system was selected following an eight month bench scale and pilot study conducted by Zimpro Passavant at its research and design facility in Wisconsin. The process diagram is shown in Figure 9.6.

The anaerobic reactor is a Zimpro Passavant Multizone-design, with biological treatment occurring in the lower, suspended growth zone, and polishing and solid-liquid separation in the upper, or fixed-film zone. The presence of the carbon enhances biological activity and increases system stability against toxics. After leachate is treated anaerobically, it passes to a circular, "nested" aeration-clarification tank. In this unit the leachate is treated with the carbon-activated sludge mixed liquor. In the anaerobic reactor the carbon is actually introduced through waste activated sludge which is returned to the anaerobic reactor. The treated effluent meets the stringent pretreatment standard before discharge to the municipal sewer

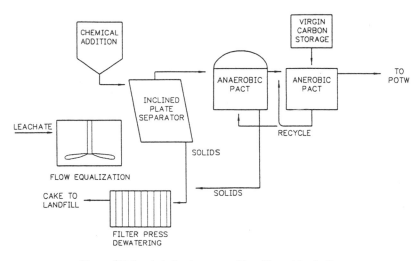

Figure 9.6 Leachate treatment at a New Hampshire facility.

system. Gas produced in the multi-zone reactor is piped to the gas-recovery plant where it is used along with the landfill gas for power generation. The sludge is dewatered in a plate and frame filter press. The cake is landfilled.

9.4.4.3 Omega Hills Landfill

The Omega Hills landfill, owned and operated by Waste Management, Inc., is located 15 miles northwest of Downtown Milwaukee, Wisconsin. The landfill leachate treatment system utilizes a unique design of anaerobic filter to pretreat 7600 m³/d (200,000 gpd) leachate generated at the landfill. The effluent is discharged into a POTW. Carter et al. (1984, 1986), and Schafer et al. (1986) provided the system design and operational data.

The treatment plant has an upflow anaerobic filter. Leachate is fed just below the media such that swirling motion provides mixing. Liquid from the top of the digester sidewall is recirculated to the inlet zone. The recirculation ratio Q_r/Q is 10. Both influent and recycle flows are heated to 35°C (95°F). Phosphoric acid is added to provide phosphorus need. The basic design features and loading rates of the filter are provided in Table 9.11. A solids contact clarifier with 9 m (30 ft) diameter and 4.9 m (16 ft) sidewater depth is used for clarification.

The treatment facility provides 80–90% BOD removal. Copper, lead and zinc removals exceed 80%. Cadmium and nickel removals are 20 and 51% respectively.

9.4.4.4 City of Sarnia Leachate Treatment Plant

The City of Sarnia Leachate Treatment Plant was constructed in 1989–90 to treat the leachate from a sanitary landfill that was established in 1971 (Lugowski et al. 1989, 1991). The leachate treatment plant is a physical-chemical and biological treatment facility. The estimated leachate generation is 91 m³/d. Since the source of leachate is an old landfill, it was difficult to treat biologically. Therefore, the process train included physical-chemical treatment followed by biological treatment.

The physical-chemical treatment system consists of preaeration column, pH adjustment, flash mixing, flocculation, and settling. A pre-oxidation stage with hydrogen peroxide as the primary oxidant is also included. Phosphoric acid is added to provide the phosphorus need. After settling, the effluent is neutralized with sulfuric acid prior to introduction into the biological treatment plant. The sludge from the settling basin is pumped into a sludge storage basin where it is allowed to concentrate further prior to dewatering by plate and frame filter presses. Approximately 0.75 m³ of sludge is produced per week at 30 to 50% solids. The supernatant is decanted and is sent to the aeration basin. The biological treatment unit consists of three aeration basins each 168 m³ in volume. The plug flow aeration basin provides 5.5 days detention time. Oxygen is supplied through coarse diffusers. The mixed liquor is settled and sludge is returned to the aeration basin. The waste activated sludge is pumped to the sludge handling area. The effluent is filtered in gravity sand filters that have air scouring and automatic backwashing. The filter backwash is returned to the aeration basin.

The effluent leaving the plant is either discharged into wetlands, or sent to post-aeration and storage lagoons. The lagoons provide detention time of four months primarily for winter months. The examination of lagoon contents showed abundance of invertebrate life, minnows, toads, frogs and snapping turtles. The effluent from the plant meets the discharge permit requirements for wetlands. Sampling and monitoring programs from wetlands have shown no deterioration in surface water quality. The effluent standard and design criteria are given in Table 9.16. The capital cost of the entire facility is $2 million, and annual O & M costs are around $100,000.

9.4.4.5 Georgswerder Landfill Germany

The Georgswerder landfill is located approximately 5 km south of the center of Hamburg, Germany. The landfill contains municipal solid wastes, war rubble, oil deposits, and semi-liquid industrial and chemical wastes. The leachate contains many metals, and toxic organics such as dioxin, benzene, phenols, and chlorinated hydrocarbons. Smith and Wilderer (1986) provide the results of bench scale and pilot plant studies conducted on actual and synthetic leachates. Process description and results are provided below.

The process train of the treatment facility included (1) pretreatment coagulation, flocculation, flotation and sedimentation, (2) two stage

TABLE 9.16. Design Criteria and Effluent Standards for Sarnia Leachate Treatment Plant.

Parameter	Design Criteria	Effluent Standards
Design capacity, m³/d (gpm)	91 (63)	—
F/M, d⁻¹	0.001–0.01	—
BOD_5, mg/L	100–1000	15
COD, mg/L	800–2000	—
TSS, mg/L	10–100	15
NH_3-N, mg/L	800–2000	10

Source: Lugowski et al. (1989,1991).

sequencing batch reactor (SBR) system (3) carbon adsorption column, and (4) float and sludge incineration.

The flotation/sedimentation process is used to separate oil, organics, and chemical precipitates from the liquid stream. Many toxic organic substances, including oil, hydrocarbons (benzene, xylene, toulene, phenol) and chlorinated compounds (dioxin, trichlorethylene, lindane) concentrate in the flotation overflow. These concentrates are incinerated. Metal precipitates are thickened, dewatered, incinerated and deposited in an abandoned salt mine.

The effluent from the flotation/sedimentation basin contained high organics. Two-stage SBR systems are used to stabilize the organics, followed by carbon adsorption for removal of refractory organics. Off gases from the biological treatment system are passed through carbon filter columns.

The SBR process was selected because of its successful application in other hazardous waste stream treatment systems. The process produced better stability and greater control compared with continuous flow process. Periodic filling and decanting of the reaction basins, and separation of the biological solids from the effluent with periodically changing environmental conditions in a controlled manner, produced enriched microbiological population that had the desired metabolic capabilities and settling characteristics. Two-stage biological treatment provided an opportunity to develop two different microbiological populations with specialized metabolic capabilities that maximized the biodegradation. A fixed-bed SBR in second stage of the treatment provided needed surface for the growth of bacterial film in relatively low substrate concentration. A silicon-membrane oxygenation system provided oxygen without formation of gas bubbles. The design and operation of both stages of the SBR reactor is given by Smith and Wilderer (1986). The final results of pilot testing are summarized in Table 9.17.

TABLE 9.17. Average Influent and Effluent Concentrations and Process Removal Efficiency of Georgswerder Treatment Facility.

| Parameter | Influent | Effluent | | Overall Efficiency, % |
		Stage 1	Stage 2	
COD, mg/L	1,170	155	75.9	93.5
TOC, mg/L	380	60.3	31.1	91.8
TOX[a], mg/L	1,085	98.8	63.8	94.1
NH$_4$-N, mg/L	300	<1	<1	99.9
NO$_3$-N, mg/L	0	175.3	172	—

[a]TOX = chlorinated hydrocarbons.
Source: Smith and Wilderer (1986).

9.4.4.6 BKK Landfill Site

Kellems et al. (1990) developed leachate treatment system for BKK Landfill located in West Covina, California. This landfill occupies a total area of 680 acres. The landfill operations commenced in 1963, when non-hazardous municipal and commercial solid wastes were accepted. The landfill accepted hazardous wastes from 1972 until 1984. Operators continued with the dumping of non-hazardous wastes on top of the hazardous disposal area. The leachate treatment plant construction and operation was part of the RCRA 1987 closure plan for the site.

The leachate collection was achieved by vertical and horizontal wells that were drilled longitudinally into the landfill lifts. The leachate had a BOD/COD ratio of 0.49, characteristic of a younger landfill. The concentration of COD, TDS, oil and grease, and phenol were 3,240, 600, 72 and 1.0 mg/L respectively. The leachate disposal options were: discharge into POTW, discharge into watercourse, or use of effluent for site irrigation on capped area.

The process train included equilization basin, chemicals nutrient adjustment, aeration basin with PAC, clarification, followed by effluent holding, then site irrigation. The sludge was dewatered by filter presses and the cake was landfilled. The facility was designed for a flow of 190 m³/d (50,000 gpd). The equilization basin had a detention time of 2.4 h, and aeration basin had HRT of 3 days. Sludge age and F/M ratio were 20 days and 0.2 per day. PAC dosage was 72.0 mg/L. The treatment facility was placed in operation in 1987 and successfully treated the leachate to a level where priority pollutant organics were below detection limit. The treated effluent was blended with surface runoff from the landfill, stored and used for landscape irrigation at the site. Due to high evapotranspiration in the area, there was zero discharge.

9.4.4.7 CID Landfill

Li et al. (1990) conducted biological treatment of hazardous waste landfill leachate in a full-scale biological sequencing batch reactor (SBR) system. These investigators studied leachate treatability from ten landfills: seven from Chemical Waste Management (CWM), Inc., containing hazardous wastes; and three from Waste Management of North America (WMNA), Inc., co-disposal MSW, and hazardous waste sites. The objective was to provide data to the U.S. EPA for hazardous waste landfill leachate standards. Part of this study involved characterization of leachates from various hazardous and co-disposal sites, as well as the evaluation of the biological leachate treatment plant at WMNA's CID landfill in Calumet City, Illinois.

The SBR facility at CID utilized process train as follows: leachate unloading area and pumps, leachate storage tanks, blending and chemical feed for pH adjustment, nutrients addition, SBR, and effluent discharge to POTW. The sludge was aerobically stabilized, thickened, dewatered by filter press, dried and placed in a secure landfill.

The SBR system at CID not only removed BOD, COD and NH_3-N efficiently, but also removed toxic organics and metals very effectively. The average removal efficiency for most of the best demonstrated available technology (BDAT) constituents was 93% as compared to 84% for COD removal, 98.7% for BOD removal, 99% for NH_3N removal and metals and inorganics removals of 56.6% and 58.1% respectively.

These investigators suggested that the removal mechanisms for volatile organics in the SBRs were most likely biological oxidation and air stripping since no volatile organics were detected in the reactor biosludge. However, as with metals and inorganics, the semivolatile organics accumulated in the biosludge in significant amounts. The two organic constituents found in the sludge with an excess concentration were paracresol and phenol.

As the solids in the sludge became more concentrated through the sludge dewatering and the drying processes, the concentration for metals and inorganics on a wet weight basis increased. However, there was a net decrease in dry weight for metals and inorganics presumably due to loss of metals and inorganics in the filtrate from the dewatering operation and in scrubber water from the dryer operation. For semivolatile organics, especially phenol and paracresol, there was a net reduction in both dry and wet weight basis during dewatering and also during drying. The loss of semivolatile organics in the dewatering process was over 98%, indicating that most organics were probably in the liquid phase, and, as moisture content of the sludge was reduced, so were the organics.

9.4.4.8 The ZenGem™ Process

Zenon Environmental, Inc. supplies a proprietary process that consists of a suspended growth activated sludge system (biological reactor) integrated with an ultrafiltration membrane system. The ultrafilter produces high quality effluent for reuse.

Janson (1992) conducted pilot testing. Many reported benefits of the process include: (1) improved effluent quality, (2) reduced sludge production; (3) improved biological degradation of retained organics, and (4) performance not affected due to sludge bulking and other upsets. Janson (1992), and Janson and Misra (1993), reported results of several pilot plant operations with this process on oily wastewaters, and industrial liquid transfer station wastes. The system successfully handled wastes containing COD

6500 mg/L, BOD 4300 mg/L, total fat, oil, and grease (TFOG) 2300 mg/L and hydrocarbon, fats, oil and grease (HFOG) 3500 mg/L. The removal efficiencies of these pollutants were 92%, 97%, 97% and 98%. The pilot test data was collected over 4 to 12 month periods. During this period the HRT varied from 0.9 to 3.7 days, and SRT of 35 to 100 days. The relative treatment cost compared favorably with conventional treatment processes. Currently, this process is being evaluated for several industrial wastes, including leachate from sanitary landfills. The process has potential for leachate treatment. Extensive technical and economic analysis must be conducted over a long time to establish performance data on leachate treatment from young and old landfills (Hare 1990; Janson 1992; Janson and Misra 1993).

9.5 SELECTION OF AN EFFECTIVE LEACHATE TREATMENT TRAIN

Many studies previously summarized clearly show that the use of physical-chemical treatment processes on leachate from young fills does not produce the degree of organic removal that can be accomplished with biological treatment processes. However, good results with physical-chemical treatment are observed with stabilized leachate collected from old fills. Similarly, good results are obtainable with leachate which has been stabilized biologically with both anaerobic and aerobic processes. Table 9.18 summarizes the proposed treatment processes for leachate as characterized by four parameters: COD/TOC and BOD/COD ratios, absolute COD concentration, and the age of the fill. These parameters must be known or estimated. Using these values, a first approximation can be made to select proper treatment processes for the removal of organic matter present in the leachate. However, if there are less than four such parameters agreeable to each other, more uncertainty is expected in the selection of the proper treatment processes. The effectiveness of various treatment processes in removing different components of leachate are presented in Table 9.19.

Based upon the literature review presented here, it is clear that the leachate characteristics will change with time, and a process train will require extensive process modifications as the landfill ages. For this reason, several investigators studied combined leachate treatment at municipal wastewater treatment plants, and concluded that combined leachate treatment at a municipal wastewater treatment plant should be the preferred method. Leachate up to 20 percent volume of municipal wastewater can be successfully treated. For this purpose, pumping of leachate, even over relatively long distances, should be given consideration.

Because leachates can have high concentrations of both organic and

TABLE 9.18. Proposed Relationship Between COD/TOC, BOD/COD, Absolute COD, and Age of Fill with Respect to Expected Efficiencies of Organic Removal from Leachate.

Character of Leachate				Effectiveness of Treatment Processes						
COD/TOC	BOD/COD	Age of Fill	COD, mg/L	Biological Treatment	Chemical Precipitation (with Lime)	Chemical Oxidation Ca(ClO)$_2$	O$_3$	Reverse Osmosis	Activated Carbon	Ion Exchange Resin
>2.8	>0.5	Young (<5 yr)	>10,000	Good	Poor	Poor	Poor	Fair	Poor	Poor
2.0–2.8	0.1–0.5	Medium (5 yr–10 yr)	500–10,000	Fair	Fair	Fair	Fair	Good	Fair	Fair
<2.0	<0.1	Old (>10 yr)	>500	Poor	Poor	Fair	Fair	Good	Good	Fair

Source: Chian and DeWalle (1977).

TABLE 9.19. Comparative Performance of Various Treatment Processes for Leachate Treatment.

Treatment Processes	Organics Young <5 yr	Organics Middle 5–10 yr	Organics Old >12 yr	Metals	VOCs	Nitrogen	Priority Pollutants	Solids	Remarks
Physical									
Natural evaporation	Good	Good	Good	Good	Good	Good	Good	Good	Applies to small flows under favorable climatic conditions
Flotation	NA	NA	NA	Fair	Fair	NA	Fair	Good	Oil, colloidal particles
Air stripping	NA	NA	NA	NA	Good	Good	Fair	NA	
Filtration	NA	NA	NA	Good	NA	NA	NA	Good	Suspended solids only
Membrane processes	Good	Good	Good	Good	Fair	Good	Good	Good	Needs pretreatment
Chemical									
Coagulation/precipitation	Poor	Fair	Poor	Good	NA	Poor	NA	Good	
Chemical oxidation	Poor	Fair	Fair	NA	Fair	NA	Good	NA	
Ion exchange	Poor	Fair	Fair	Good	NA	Fair	NA	Good	Pretreatment is required
Carbon absorption	Poor	Fair	Good	NA	Good	NA	Good	NA	Pretreatment is required
Biological									
Aerobic suspended growth	Good	Fair	Poor	Good	Good	Fair	Fair	Fair	
Aerobic fixed film	Good	Fair	Poor	Good	Good	Fair	Fair	Fair	
Anaerobic suspended growth	Good	Fair	Poor	Good	Good	Fair	Fair	Fair	
Anaerobic fixed film	Good	Fair	Poor	Good	Good	Fair	Fair	Fair	

NA = not applicable.

inorganic contaminants, treatment of leachate must utilize integration of the basic methodologies into a systematic approach. The design of such a system requires considerations not only to the volume and quality of leachate to be treated, but also to the changes in leachate quality that may occur over time as the landfill ages. Several leachate treatment system options must be considered. Among these are:

(1) Leachate pretreatment for discharge into POTWs
(2) Leachate pretreatment for zero discharge
(3) Leachate treatment for disposal into receiving waters

All these systems are discussed in the following pages.

9.5.1 LEACHATE PRETREATMENT FOR DISCHARGE INTO POTW

Under the NPDES permitting program of the Clean Water Act, pretreatment standards are being developed for all pollutants that "interfere with, pass through, concentrate in sludge, or otherwise are incompatible" with publicly owned treatment works (POTW).

Under the pretreatment regulations, two types of federal pretreatment standards are established: (1) prohibited discharges and (2) categorical standards. Prohibited discharges are those that cause fire or explosion hazard, corrosion, obstruction, slug discharges, and heat discharge to sewers or POTWs. The categorical standards are developed for those pollutants that are incompatible, that is, those that interfere with the operation of, pass through, or concentrate into the sludge and other residues from POTWs.

The substances considered for categorical standards are those for which there is substantial evidence of carcinogenicity, mutagenicity, and/or teratogenicity; substances structurally similar to aforementioned compounds; and substances known to have toxic effects on human beings or aquatic organisms at sufficiently high concentration, and which are present in the industrial effluents. There are many specific elements or compounds that have been identified as priority pollutants. These include metals, organics, cyanides, and asbestos. The updated list may be obtained from U.S. EPA.

The NPDES permitting and pretreatment programs as they exist today are part of a very complex regulatory scheme. Categorical standards and guidelines for pretreatment and effluent discharges are being established for toxic, conventional, and nonconventional pollutants.

Landfill leachate has high concentrations of organic and inorganic pollutants that change with time as landfill ages. Combined leachate treatment in municipal wastewater treatment plants has been extensively investigated at bench scale, pilot plant, and full scale facilities. It has been demonstrated that combined treatment should be considered as the preferred method of treatment if a municipal sewer system is conveniently available. Many pretreatment systems have also been investigated. Among these are conventional suspended growth, and attached growth biological reactor systems. The effluent is discharged into the sewer, and thickened sludge or sludge cake is returned into the landfill.

The most cost-effective pretreatment method of landfill leachate is aerated lagoon followed by stabilization pond. Properly designed units with impervious liners are proven technology, if land area and site conditions permit. A diffused or mechanical aeration system with aeration period of 2–5 days should be installed to provide mixing, and satisfy the oxygen requirements. Lime or coagulant addition will enhance heavy metal precipitation. Stabilization ponds having detention time of 4–10 days will provide sedimentation, and will serve as a storage basin. Proven design techniques are available for designing these facilities for all geographical areas including cold weather conditions. Table 9.20 provides a comparative rating of several pretreatment systems for sanitary landfill leachate.

9.5.2 LEACHATE TREATMENT WITH ZERO DISCHARGE

If the leachate quantity is small it is best to utilize a process train that provides potential for zero discharge. Among these applicable methods are (1) natural or solar evaporation, and (2) land application.

9.5.2.1 Natural or Solar Evaporation

The natural or solar evaporation of industrial waste brines has been practiced extensively where climatic conditions permit. The process involves large impoundments with no discharge. The amount of evaporation from water surface depends on temperature, wind velocity, and humidity. There are substantial variations in the average daily evaporation rate from month to month, and year to year. However, depending on the climatic conditions, large impoundment may be necessary if precipitation exceeds evaporation over several months. Therefore, consideration must be given to net evaporation, storage requirements, seepage and percolation, and groundwater protection (Davies 1973).

For leachate treatment, the natural evaporation process can be easily adapted as land area is generally available on the property. The area requirements can be reduced by leachate recirculation through the landfill during the adverse climatic conditions. The recirculation will provide temporary storage in the landfill. The recirculation will also assist in stabilization of leachate and the landfill. Design procedure for solar evaporation ponds are well established. Some of the techniques discussed in

TABLE 9.20. Comparison of Various Pretreatment Systems for Sanitary Landfill Leachate.

Pretreatment Systems	Land Area Needed	Construction Cost	Operation and Maintenance Requirements	Performance Level	Ease of Upgrading With Age of Fill
Aerated lagoon and stabilization	High	Low	Low	Medium	Medium[a]
Anaerobic digester	Medium	High	Medium	Low	Low
Extended aeration	Low	High	High	High	Good[b]
Trickling filter	Medium	High	Medium	High	Low
RBC	Low	High	High	High	Low
Physical chemical	Low	High	High	High	High

[a]Chemicals may be used to enhance precipitation.
[b]Can be converted into a physical-chemical treatment system.

Chapter 5 for leachate generation can be easily adapted to design an evaporation facility.

9.5.2.2 Land Application

Land treatment of municipal wastewater is a proven technology. It provides the possibility for nutrient reuse, and produces effluent of a high quality. Land treatment of leachate has been evaluated on an experimental basis. Early investigations have shown that leachate can be effectively treated by land application without any adverse contamination of soil. Pretreatment of leachate as proposed for POTW discharge would enhance the applicability of land treatment. Nordstedt (1975) applied leachate over pasture grass and found no change in soil quality.

Land treatment of municipal and industrial wastewaters includes the use of plants, land surface, and the soil matrix to remove various constituents in the wastewater by many physical, chemical, and biological means. The objectives of land treatment include irrigation, nutrient reuse, crop production, recharge of groundwater, and water reclamation for reuse. There are three basic methods of land application: (1) slow-rate irrigation, (2) rapid infiltration-percolation, and (3) overland flow. Each method can produce renovated water of different quality, can be adapted to different site conditions, and can satisfy different overall objectives.

9.5.2.2.1 Slow-Rate Irrigation

Irrigation is the most widely used form of land treatment system. The wastewater is applied to the crops or vegetation either by sprinkling or by surface techniques. In this process, surface runoff is not allowed. A large portion of water is lost by evapotranspiration, whereas some water may reach the groundwater table. Groundwater quality criteria may be a limiting factor for system selection. Some factors that are given consideration in the design and selection of an irrigation method are (1) availability of suitable site, (2) type of wastewater and pretreatment, (3) climatic conditions and storage needed, (4) soil type and organic or hydraulic loading rate, (5) crop production, (6) distribution methods, (7) application cycle, and (8) ground and surface water pollution (Qasim 1994).

9.5.2.2.2 Rapid Infiltration-Percolation

Rapid infiltration-percolation differs from the irrigation method in that wastewater percolates through the soils and treated effluent reaches the groundwater. This method is not suitable for leachate treatment. Plants are not used for evapotranspiration as in irrigation systems. The objectives of rapid infiltration-percolation are: (1) groundwater recharge, (2) natural treatment followed by withdrawal by pumping or underdrain systems for recovery of treated water, and (3) natural treatment with groundwater moving vertically and laterally in the soil, and recharging a surface water course.

9.5.2.2.3 Overland Flow

In an overland flow system, the wastewater is applied over the upper reaches of the sloped terraces and allowed to flow overland and is collected at the toe of the slopes. The collected effluent can be either reused or discharged into the receiving water. Biochemical oxidation, sedimentation, filtration, and chemical adsorption are the primary mechanisms for removal of contaminants. Nitrogen removal is achieved through denitrification.

9.5.2.2.4 Other Systems

Wastewater treatment and beneficial use of nutrients contained in the wastewater may also be achieved through silviculture, aquaculture, and application in existing or artificial marshlands and wetlands. Various combinations of plants and animals from microscopic organisms to fish and water hyacinths have been attempted with some success for water quality improvements and other beneficial uses.

9.5.2.2.5 Design Considerations

Design and selection of land treatment systems is complex. Many engineering, environmental, health, physical and social science, legal, and economic factors warrant detailed investigation. Accumulation of toxic chemicals in soils and food chain, groundwater contamination, and odor problems have received wide concern. Table 9.21 provides a summary of many factors that must be considered in the evaluation and selection of a land treatment system. In all respects, the land treatment of leachate must comply with the regulations promulgated under the Resource Conservation and Recovery Act (as amended) to protect human health and the environment from the improper management of landfill leachate.

9.5.3 LEACHATE TREATMENT FOR DISPOSAL IN NATURAL WATER SYSTEMS

Under the Clean Water Act requirements, effluent standards are imposed upon individual point source discharges. These requirements are

TABLE 9.21. General Design Considerations That Must Be Studied for Design and Selection of Land Treatment Methods.

Wastewater Characteristics	Climate	Geology	Soils	Plant Cover	Topography	Application	Environmental
Volume	Total precipitation and frequency	Groundwater	Type	Indigenous to region	Slope	Method	Others
Constituent load BOD Nitrogen Phosphorus Heavy Metals Other toxics	Evapotranspiration Temperature Growing season Occurrence and depth of frozen ground Storage requirement Wind velocity and direction	Seasonal depth Quality Points of discharge Bedrock Type Depth Permeability	Gradation Infiltration/permeability Type and quantity of clay Cation exchange capacity Phosphorus adsorption potential Heavy-metal adsorption potential pH Organic content	Nutrient removal capability Toxicity levels Moisture and shade tolerance Marketability	Aspect of slope Erosion hazard Crop and farm management	Type of equipment Application rate Types of drainage	Legal Socio-economic

Source: Adapted from Qasim (1994).

287

backed up by a permit program which describes the effluent limits, monitoring requirements, and compliance schedule. The leachate treatment system must meet the level of effluent quality established for the discharge. It is well known that the leachate characteristics will change with time, and the process train will need extensive process modifications as the landfill ages to meet the effluent limitations. For this reason a treatment system must be developed that can meet the current limitations, and can be modified with time to accommodate the changing characteristics of the leachate. Several process trains can be developed and modified to meet the requirements. To develop the best possible process train, a designer must evaluate many factors that are generally related to operation and maintenance, process efficiency under variable flow and changing influent quality conditions, and environmental constraints. In Table 9.22 various factors that are considered important in selection of process trains for complete leachate-treatment are evaluated. Examples of several process trains are presented below. The reasons for selecting them, and how they should be modified to effectively treat the leachate as its quality changes with time, are also discussed.

9.5.3.1 Process Train 1—Two-Stage Suspended Growth Reactors

Biological treatment is quite effective in removing organics from leachate of young landfills. However, a single-stage activated sludge process is not effective in removing residual refractory organics remaining after stabilization of various biodegradable organic material. The prime reason is that microorganisms which can degrade the refractory organic material are not fully developed in a single-stage activated sludge system. The second-stage of a two-stage activated sludge system can maintain microorganisms acclimated to the specific organic matter in the separate aeration system. These microorganisms will effectively remove the refractory organics that exert COD in the effluent. A process train designed on this concept is shown in Figure 9.7. For young and middle-age landfills, this process train will utilize an anaerobic reactor that will also act as an equalization basin [Figure 9.7(a)]. The leachate treated in the first-stage activated sludge process will be further treated in the second-stage activated sludge process. In the first aeration basin, a large portion of the BOD will be stabilized. The refractory organics will be removed in the second stage aeration basin. If necessary, powdered activated carbon may be added into the second-stage aeration basin to enhance the removal of the refractory organics. It is expected that complete nitrification will be achieved in the second-stage activated sludge system. The denitrification unit is employed after the second stage aeration system to remove nitrogen content.

As the landfill ages and the chemical characteristics of the leachate change, the second-stage suspended growth reactor will be modified into a chemical/physical treatment facility, and an ammonia stripping tower will be added. The denitrification facility will be converted for recarbonation and pH adjustment of the effluent from the ammonia stripping tower. This modification is shown in Figure 9.7(b). The sludge generated will be returned to the landfill.

This process train, with proper modifications in time, will provide effective treatment of biodegradable and refractory organics, heavy metals, and nitrogen and phosphorous over the entire life of the fill. This process train will not provide removal of dissolved salts.

9.5.3.2 Process Train 2—Sequencing Batch Reactor, (SBR), Biologically Active Carbon (BAC) Filter, and Denitrification

A potentially attractive process train is to utilize anaerobic digester/equalization basin, coagulation flocculation and sedimentation for pretreatment, SBR, BAC, bed, and denitrification. This process train initially will provide an effective method to treat the leachate from young and middle-age landfills.

The anaerobic digester will provide equalization. The flocculation/precipitation facility will remove heavy metals. Phosphorous addition prior to the SBR system may be needed. The SBR system will provide enriched microbiological population that is expected to have desired metabolic capabilities, stability, and settling characteristics. The biodegradable organics as well as refractory organics will be stabilized. The biologically active carbon will provide desired biological film to metabolize the refractory organics present in relatively low concentrations, and provide nitrification of remaining ammonia. Nitrate nitrogen will be removed in a denitrification facility. The process train is shown in Figure 9.8(a). The facility will provide removal of biodegradable and refractory organics, heavy metals and nutrients.

With time, as the landfill ages, the leachate quality will change. The facility may be upgraded by addition of granular activated carbon columns. The upgraded facility is shown in Figure 9.8(b).

9.5.3.3 Process Train 3—Biological Treatment and Reverse Osmosis

A treatment system to also remove inorganic solids from leachate will utilize demineralization. The proposed process train will include anaerobic digester/equalization basin, suspended growth biological waste treatment system, partial softening, filtration and RO. The anaerobic digester will provide solubilization of organics and flow equalization. The extended

TABLE 9.22. Factors Considered in Selection of Complete Treatment of Landfill Leachate.

Treatment Processes	Land Requirement	Adverse Climatic Conditions	Ability to Handle Flow Variations	Ability to Handle Influent Quality Variations	Reliability of the Process	Ease of Operation & Maintenance	Ease of Upgrading Process Change	Occupational Hazards	Air Pollution	Waste Products
Physical										
Flotation	Small	Low	Poor	Poor	Fair	Poor	Fair	Chemicals, pressure vessel	VOC	Scum
Air stripping	Small	High	Fair	Fair	Good	Fair	Poor		VOC Ammonia	Ammonia
Filtration	Small	Low	Poor	Fair	Good	Good	Good			Backwash
Membrane processes	Small	Medium	Poor	Good	Good	Fair	Fair	Pressure vessel		Brine
Chemical										
Coagulation/ precipitation	Medium	Low	Good	Good	Good	Fair	Good	Chemicals		Sludge
Chemical oxidation	Small	Medium	Fair	Fair	Fair	Poor	Fair	Chemicals	Chemical odor	
Ion exchange	Small	Low	Poor	Fair	Fair	Fair	Fair	Chemicals		Brine

TABLE 9.22. (continued).

Treatment Processes	Land Requirement	Adverse Climatic Conditions	Ability to Handle Flow Variations	Ability to Handle Influent Quality Variations	Reliability of the Process	Ease of Operation & Maintenance	Ease of Upgrading Process Change	Occupational Hazards	Air Pollution	Waste Products
Carbon adsorption	Small	Low	Poor	Poor	Good	Fair	Fair	Fire explosion	Regenerant gases	Spent carbon
Biological										
Aerobic suspended growth	Medium	Medium	Good	Fair	Good	Good	Good		VOC	Sludge
Aerobic fixed film	Large	Low	Fair	Good	Good	Good	Poor		VOC	Sludge
Anaerobic suspended growth	Large	High	Good	Fair	Fair	Fair	Good	Explosion, Fire		Sludge
Anaerobic fixed growth	Medium	High	Fair	Good	Good	Good	Poor	Explosion, Fire		Sludge
Sequencing batch reactor	Medium	Medium	Fair	Good	Fair	Poor	Poor		VOC	Sludge

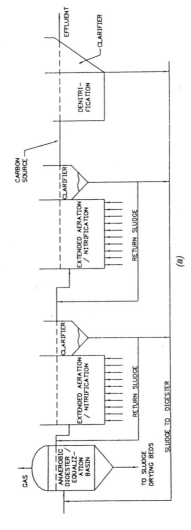

Figure 9.7 Process Train 1 for complete leachate treatment: (a) young to medium age landfill.

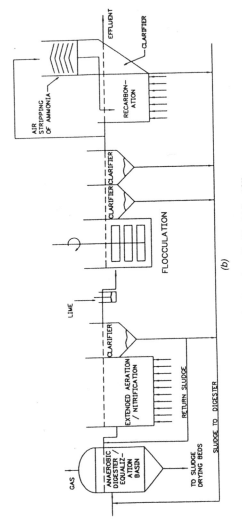

Figure 9.7 (continued) (b) old landfill.

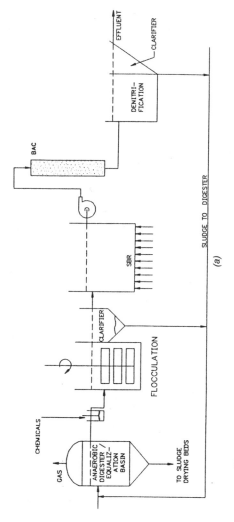

Figure 9.8 Process Train 2 for complete leachate treatment: (a) young to medium age landfill.

(a)

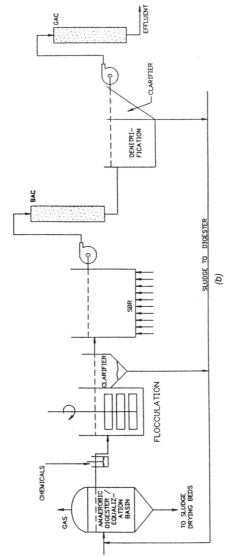

Figure 9.8 (continued) (b) old landfill.

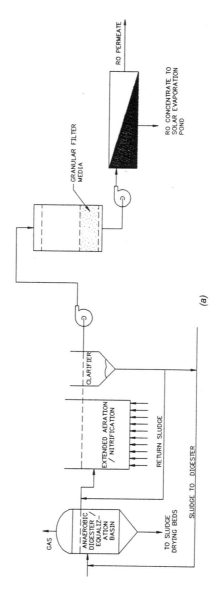

Figure 9.9 Process Train 3 complete leachate treatment: (a) young to medium age landfill.

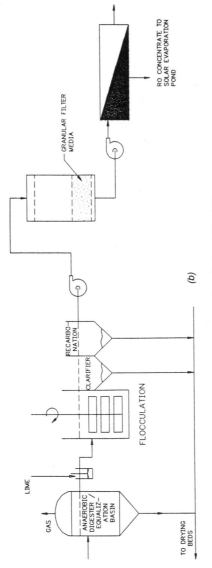

Figure 9.9 (continued) (b) old landfill.

aeration system will provide biological decomposition of organic matter. The effluent will be polished in a gravity filter followed by demineralization in a RO unit. High quality effluent produced by the system will be low in organics, and dissolved salts. The effluent quality will permit reuse for any desired application. The brine and sludge may be returned to the landfill, or evaporated in a solar pond. The process train is shown in Figure 9.9(a). As the landfill ages, the biological treatment process will be replaced by a coagulation precipitation facility, followed by recarbonation, filtration, and RO system. The upgraded facility is shown in Figure 9.9(b).

9.5.3.4 Process Train 4—Microfiltration/RO

Zenon Environmental, Inc. has developed a process that holds promise for treatment of low to medium flow landfill leachate. The system is undergoing extensive pilot-plant, and full-scale treatment of leachate from municipal and industrial-hazardous landfills. The advantages of the system are (a) high quality product water, and (b) process works equally well for young, medium age and old landfills.

The process train is shown in Figure 9.10. The two stage process utilizes (a) precipitation and microfiltration for removal of toxic metals and suspended solids, and (b) concentration of residual organics by reverse osmosis. The first step precipitation/microfiltration system provides a simple pretreatment stage for RO.

Chemical precipitation removes metals, calcium and magnesium. The permeate from microfiltration membrane will have low suspended solids (less than 10 mg/L). This permeate is effectively treated by the RO unit. At this time Zenon Environmental, Inc. is conducting extensive full scale testing with landfill leachates to develop design and operational procedures,

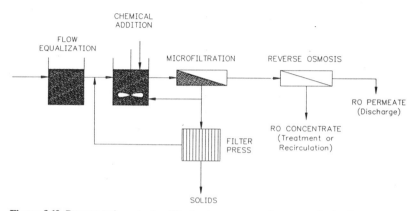

Figure 9.10 Process train and microfiltration/reverse osmosis process for leachate treatment.

and to reduce membrane fouling. The process train remains effective as landfill ages.

9.6 LEACHATE TREATMENT COSTS

Developing reliable leachate treatment costs is difficult, as many cost components are involved. Associated cost factors are (1) leachate collection system, possibly including an impermeable barrier or lines beneath the fill, drains and sumps, (2) pumping equipment and (3) treatment facility. Tolman et al. (1978), Baldwin (1979), Lu et al. (1984), and Matrecon, Inc. (1988), have provided cost data on various components of leachate collection and treatment systems. It has been estimated that the leachate containment and collection system may increase the costs 30 to 40 percent over an equivalent landfill without containment and collection systems (Baldwin, 1979). Straub and Lynch (1982) estimated that the total cost of leachate collection, treatment, and handling may increase the cost of solid waste disposal by 25–30%.

Chian and DeWalle (1977) developed unit costs of several physical, chemical and biological treatment processes. The cost of leachate treatment will vary greatly as leachate quantity and quality changes with time, and the facility may need extensive upgrading. It is therefore not possible to generalize the cost of leachate treatment. Leachate recirculation has been reported to reduce the treatment cost since concentration of organics are reduced, and landfill can act as an equalization basin to maintain a nearly constant flow. Straub (1980), provided comparative annual treatment costs for leachate from recirculated and non-recirculated landfills. A net saving of 20% for large, and 40% for small landfills can be achieved by recirculation. In these cost savings, the construction and operation costs of leachate recycling systems were included. Other earlier studies also provided many treatment costs and related information. Steiner et al. (1979) reported that energy cost for activated sludge and ammonia stripping consumed 50% of total treatment cost. Vydra and Grimm (1976), in one instance, estimated the capital cost of leachate collection and treatment amounted to about 10% of the total construction cost of the landfill. This cost was kept low because of the native clay soil for the liner. The cost of installing an impermeable liner could more than double the capital cost figure of the leachate collection and treatment system.

Chiang, Patel and Associates, Inc. has developed capital, and operation and maintenance cost estimates for various treatment processes applicable to leachate treatment. These costs are developed for three average flows: 20, 75 and 380 L/min (5, 20 and 100 gpm). Several simplifying assumptions have been made to estimate the costs. These are:

TABLE 9.23. Capital, Operation and Maintenance Costs of Leachate Treatment Processes, (1994 Dollars).

Treatment processes	Capital cost thousands of $			Annual O & M cost thousands of $			Total Unit cost[a] $/m³		
	20, L/min (5 gpm)	75, L/min (20 gpm)	380, L/min (100 gpm)	20, L/min (5 gpm)	75, L/min (20 gpm)	380, L/min (100 gpm)	20, L/min (5 gpm)	75, L/min (20 gpm)	380, L/min (100 gpm)
Evaporation pond	100	300	600	10	15	20	1.78	1.04	0.36
Stabilization pond	60	200	350	10	15	20	1.45	0.82	0.25
Aerated lagoon	75	250	450	15	20	40	2.05	1.06	0.40
Extended aeration with clarifier	120	400	700	15	30	60	2.42	1.65	0.61
Trickling filter with clarifier	85	275	500	15	25	40	2.13	1.24	0.42
RBC with clarifier	100	300	600	15	30	60	2.26	1.42	0.56
SBR	85	275	500	15	30	60	2.13	1.37	0.52
Biological denitrification[b]	30	100	200	15	20	40	1.68	0.73	0.29
Anaerobic digester[c]	40	125	250	15	20	40	1.76	0.78	0.31
Anaerobic filter[c]	40	125	250	15	20	40	1.76	0.78	0.31

TABLE 9.23. (continued).

Treatment processes	Capital cost thousands of $			Annual O & M cost thousands of $			Total Unit cost[a] $/m³		
	20, L/min (5 gpm)	75, L/min (20 gpm)	380, L/min (100 gpm)	20, L/min (5 gpm)	75, L/min (20 gpm)	380, L/min (100 gpm)	20, L/min (5 gpm)	75, L/min (20 gpm)	380, L/min (100 gpm)
Coagulation/precipitation,[d] with clarifier	40	125	250	15	30	60	1.76	1.04	0.41
Ammonia stripping[e]	25	80	150	10	15	30	1.16	0.56	0.22
Gravity filter[b]	30	100	200	15	20	40	1.68	0.73	0.29
PAC feed, dose 500[b] mg/L	25	80	150	15	20	40	1.63	0.68	0.27
GAC[f]	60	300	350	20	40	80	2.40	1.68	0.55
RO[f]	70	225	400	20	40	80	2.48	1.51	0.58
Microfiltration with RO[f]	75	250	450	25	50	90	3.00	1.82	0.65

[a]interest rate 6%, service life 20 years, capital recovery factor 0.08718.
[b]Follows biological treatment.
[c]50–70% organics removal.
[d]30–50% organics removal.
[e]High pH, only NH_3-N removal.
[f]Follows biological treatment or coagulation/precipitation with filtration.
Source: Treatment cost estimates are developed by Chiang, Patel and Associates, Inc., interest rate–6% and service life 20 years. All costs are in 1994 dollars.

(1) During low flows or no flows the effluent is recirculated through the fill to maintain a flow through treatment plant for sustaining the biological growth. During the high flows the leachate will be pumped to maintain the design capacity through the treatment plant. The landfill, therefore, will provide some flow equalization.

(2) The cost of leachate collection and recirculation systems are not included in treatment costs. These costs are provided separately in Chapter 8.

(3) Land cost is not included. It is assumed that sufficient land is available at the landfill site for construction of the treatment facility.

(4) A service life of 20 years and interest rate of 6% are assumed.

(5) As the landfill ages and the leachate quality changes, process modifications will be necessary. The costs of future plant upgrading are not included in these cost estimates.

(6) Average leachate quality assumed for these cost estimates are: BOD_5 = 4000 mg/L, COD = 8000 mg/L, TSS = 100 mg/L, Alkalinity = 6000 mg/L as $CaCO_3$, TDS = 10,000 mg/L, and pH = 5.8

(7) Evaporation pond costs are based on net annual evaporation rate of 30 cm (12 in), and no net evaporation for three consecutive months.

(8) To achieve a desired level of treatment for a given influent quality, proper process selection should be made to develop the process train.

The capital, operation and maintenance costs of many treatment processes are summarized in Table 9.23.

9.7 REFERENCES

Anonymous. 1992. "Landfill Leachates: Forward-Looking System at Turnkey," *Reactor,* Zempro Passavant Environmental System, Inc., 73:4–5.

Baldwin, V. A. 1979. "Environmental and Resource Conservation Considerations of Steel Industry Solid Waste," EPA-600/2-78-074, US EPA, Cincinnati, Ohio.

Barchyn, P. 1984. "Feasibility of Treating Hartland Road Landfill Leachate at Central Saanich Pollution Centre," Unpublished report by Capital Regional District, Victoria, British Columbia.

Berkowitz, J. B., J. T. Funkhouser and J. I. Stevens. 1978a. *Unit Operations for Treatment of Hazardous Industrial Wastes,* Park Ridge, New Jersey: Noyes Data Corporation.

Berkowitz, J. B., J. T. Funkhouser and J. I. Stevens. 1978b. *Physical, Chemical, and Biological Treatment Techniques for Industrial Wastes,* Vol. I, U.S. Environmental Protection Agency, NTIS PB-275 054/5GI.

Berkowitz, J. B., J. T. Funkhouser and J. I. Stevens. 1978c. *Physical, Chemical, and Biological Treatment Techniques for Industrial Wastes,* Vol. II, U.S. Environmental Protection Agency, NTIS PB-275 287/1GI.

Birkbeck, A. E. 1984. "Characterization of Leachate from a Landfill on a Peat Bog," *Proceedings 6th National Conference on Waste Management in Canada,* Vancouver, B. C.

Boyle, W. C. and R. K. Ham. 1974. "Treatability of Leachate from Sanitary Landfills," *Journal Water Pollution Control Federation,* 46(5):860–872.

Burchinal, J. C. 1970. "Microbiology and Acid Production in Sanitary Landfills," A Summary Report, U.S. Department of Health, Solid Waste Program Grant No. UI 00529.

Carter, J. L., G. M. Curran, P. E. Schafer, R. T. Janeshek, and G. C. Woelfel. 1984. "A New Type of Anaerobic Design for Energy Recovery and Treatment of Leachate Wastes," *Proceedings of the 39th Purdue Industrial Waste Conference,* Purdue University, West Lafayette, Indiana, pp. 369–376.

Carter, J. L., P. E. Schafer, R. T. Janeshek and G. C. Woelfel. 1986. "Effect of Alkalinity and Hardness on Anaerobic Digestion of Landfill Leachate," *Proceedings of the 40th Purdue Industrial Waste Conference,* Purdue University, West Lafayette, Indiana, pp. 621–630.

Chian, E. S. K. and F. B. DeWalle. 1976. "Sanitary Landfill Leachates and Their Treatment," *Journal of the Environmental Engineering Division,* ASCE, 102(EE2):411–431.

Chian, E. S. K. and F. B. DeWalle. 1977. "Evaluation of Leachate Treatment, Volume II, Biological and Physical-Chemical Processes," EPA-600/2-77-186b, US EPA, Cincinnati, Ohio.

Chian, E. S. K. and H. H. P. Fang. 1973. "Evaluation of New Reverse Osmosis Membranes for the Separation of Toxic Compounds from Water," American Institute of Chemical Engineers Symposium Series, 71(145):497–507.

Chian, E. S. K., F. G. Pohland, K. C. Chang, and S. R. Harper. 1985. "Leachate Generation and Control at Landfill Disposal Sites," *Proc. New Directions and Research in Waste Treatment and Residuals Management,* University of British Columbia, pp. 14–30.

Cook, E. N. and E. G. Foree. 1974. "Aerobic Biostabilization of Sanitary Landfill Leachate," *Journal Water Pollution Control Federation,* 46:380–392.

Christensen, T. H., R. Cossu and R. Stegmann. 1992. *Landfilling of Waste Leachate,* London: Elsevier Applied Science.

Culp, G. L. and N. F. Heim. 1978. "Performance Evaluation and Troubleshooting at Municipal Wastewater Treatment Facilities," No. 16-EPA-430/9-78-001, US EPA, Washington, D.C.

Davies, R. 1973. "The Hydrogeologist in Action," *Water and Pollution Control,* 111(12):21–25.

DeWalle, F. B. and E. S. K. Chian. 1974a. "The Kinetics of Formation of Humic Substances in Activated Sludge Systems and Their Effect on Flocculation," *Biotechnology and Engineering,* 14(7):739–755.

DeWalle, F. B. and E. S. K. Chian. 1974b. "Removal of Organic Matter by Activated Carbon Columns," *Journal of the Environmental Engineering Division,* ASCE, 100(EE5):1089–1104.

DeWalle, F. B. and E. S. K. Chian. 1975. "Treatment of Leachate with Anaerobic Filter," presented at the First Chemical Congress of North America, held at Mexico City, Mexico.

Dzombak, D. A., K. M. Langnese, D. B. Spengel and R. G. Luthy. "Comparison of Activated Sludge and RBC Treatment of Leachate From a Solid Waste Landfill,"

Proceedings of 1990 WPCF National Specialty Conference on Water Quality Management of Landfills, Chicago, Ill., July 15–18, pp. (4-39)–(4-58).

Ehrig, H. J. 1985. "Biological Treatment of Sanitary Landfill Leachate with Special Aspects on High Ammonia Concentration," *Proc. New Directions and Research in Waste Treatment and Residuals Management,* University of British Columbia, pp. 232–248.

Fillos, G., P. J. Gleason and B. Liebowitz. 1990. "Leachate Treatment at the Fresh Kills Landfill," *Proceedings of 1990 WPCF National Specialty Conference on Water Quality Management of Landfills,* Chicago, Illinois, July 15–18, pp. (7-12)–(7-34).

Foree, E. G. and V. M. Reid. 1973. "Anaerobic Biological Stabilization of Sanitary Landfill Leachate," Technical Report TR 65-73-CE17, Department of Civil Engineering, University of Kentucky, Lexington, Kentucky.

Gaudy, A. F., A. F. Rozich and S. Garniewski. 1986. "Treatability Study of High Strength Landfill Leachate," *Proceedings of the 41st Industrial Waste Conference,* Purdue University, West Lafayette, Indiana, 627–638.

Graham, D. W. and D. S. Mavinic. 1979. "Biological-Chemical Treatment of Leachate,"*Proceedings of the American Society of Civil Engineers National Conference on Environmental Engineering,* San Francisco, pp. 291–298.

Hare, R. W. 1990. "Membrane Enhanced Biological Treatment of Oily Wastewater," Paper Presented at 1990 WPCF Conference, Wash., D.C.

Henry, J. G. 1985. "New Developments in Landfill Leachate Treatment," *Proceedings of International Conference on New Directions and Research in Waste Treatment and Residuals Management,* University of British Columbia, pp. 1–139.

Ho, S., W. C. Boyle and R. K. Ham. 1974. "Chemical Treatment of Leachates from Sanitary Landfills," *Journal Water Pollution Control Federation,* 46(7):1776–1791.

Janson, A. F. 1992. "The ZenoGem™ Process for Industrial Wastewater Treatment," Paper presented at 39th Ontario Conference on the Environment, Toronto, Ontario.

Janson, A. F. and P. N. Mishra. 1993. "A Novel and Effective Industrial Wastewater Treatment Process Integrating Biological Treatment and Ultrafiltration," *Environmental Science and Engineering,* January, pp. 76–77.

Karr, P. R. III. 1972. "Treatment of Leachate from Sanitary Landfills," Special Research Problem, School of Civil Engineering, Georgia Institute of Technology, Atlanta, Georgia.

Kellems, B. L., R. G. Schilcher, S. F. McShane. 1990. "Treatment of Leachate from Hazardous Waste Landfills, Three Case Studies," *Proceedings of 1990 WPCF National Specialty Conference on Water Quality Management of Leachate,* Chicago, Ill., July 15–18, pp. (3-34)–(3-53).

Kelly, H. G. 1987. "Pilot Testing for Combined Treatment of Leachate from a Domestic Waste Landfill Site," *Journal Water Pollution Control Federation,* 59(5):254–261.

Keenan, J. D., R. L. Steiner and A. A. Fungaroli. 1983. "Chemical-Physical Leachate Treatment," *Journal of Environmental Engineering Division,* ASCE, 109(6):1371–1384.

Keenan, J. D., R. L. Steiner and A. A. Fungaroli. 1984. "Landfill Leachate Treatment," *Journal Water Pollution Control Federation,* 56(1):27–33.

Kinman, R. N., J. W. Stamm and D. L. Nutine. 1985. "Reverse Osmosis Membrane Fouling by Sanitary Landfill Leachate and Copper Plating Solutions," *Proceedings of the 40th Purdue Industrial Waste Conference,* Purdue University, West Lafayette, Indiana, pp. 467–475.

Kinman, R. N. and D. L. Nutine. 1990. "Reverse Osmosis Treatment of Landfill Leachate," *Proceedings of the 45th Purdue Industrial Waste Conference,* Purdue University, West Lafayette, Indiana, pp. 617–622.

Kipling, J. J. 1965. *Adsorption from Solution of Non Electrolytes,* New York, New York: Academic Press.

Krug, T. A. and S. McDougal. 1988. "Preliminary Assessment of a Microfiltration/Reverse Osmosis Process for the Treatment of Landfill Leachate," *Proceedings of the 43rd Purdue Industrial Waste Conference,* Purdue University, West Lafayette, Indiana, pp. 185–193.

Leckie, J. O., J. G. Pacey and C. Halvadakis. 1975. "Acceleration Refuse Stabilization Through Controlled Moisture Application," Paper Presented in the Second Annual National Environmental Engineering Research Development and Design, Gainsville, Florida.

Leckie, J. O., J. G. Pacey and C. Halvadakis. 1979. "Landfill Management with Moisture Control," *Journal Environmental Engineering Division,* ASCE, 105(EE2): 337–355.

Lee, C. J. 1979. "Treatment of Municipal Landfill Leachate," M. S. Thesis, University of British Columbia, Vancouver, Canada.

Li, A., J. Arand, B. Liu and J. Urek. 1990. "Biological Treatment of Hazardous Waste Landfill Leachate: Fate of Toxic Organics and Metals in Full-Scale Biological Sequencing Batch Reactor (SBR)", *Proceedings of the 1990 WPCF National Specialty Conference on Water Quality Management of Leachate,* Chicago, Ill., July 15–18, pp. (4–59), (4–12).

Lu, J. S. C., B. Eichenberger and R. L. Stearns. 1984. "Production and Management of Leachate from Municipal Landfills: Summary and Assessment," EPA-600/2-84-092, Cincinnati, Ohio: US EPA Municipal Environmental Laboratory.

Lugowski, A., D. Haycock, R. Poisson, S. Beszedits. 1989. "Biological Treatment of Landfill Leachate," *Proceedings of the 44th Industrial Waste Conference,* Purdue University, West Lafayette, Indiana, pp. 565–572.

Lugowski, A., and R. Poisson. 1991. "Sarnia Leachate–Treatment Plant–A First of Its Kind in Ontario," *Environmental Science and Engineering,* May, pp. 46–48.

Maris, P. J., D. W. Harrington and F. E. Mosey. 1985. "Treatment of Landfill Leachate Management Operations," *Proc. New Directions and Research in Waste Treatment and Residuals Management,* University of British Columbia, pp. 280–296.

Martin, E. G. and J. H. Johnson. 1987. *Hazardous Waste Management Engineering,* New York: Van Nostrand Reinhold Company.

Matrecon, Inc. 1988. "Lining of Waste Impoundment and Other Facilities," EPA/600/2-88/052, Risk Reduction Engineering Laboratory, Office of Research and Development, U.S. Environmental Protection Agency, Cincinnati, Ohio.

McClintock, S. A., C. W. Randall, D. C. Marickovich and G. Wang. 1990. "Biological Nitrification and Denitrification Process for Treatment of Municipal Landfill Leachate," *Proceedings of the 1990 WPCF National Specialty Conference on Water Quality Management of Leachate,* Chicago, Ill., July 15–18, pp. (4-19)–(4-38).

Mavinic, D. S. 1984. "Kinetics of Carbon Utilization in Treatment of Leachate," *Water Research,* 18(10):1279–1284.

Memoin, E. P., and D. S. Mavinic. 1981. "Nutrient Requirements of Aerobic Biostabilization of Landfill Leachate," *Proceedings of the 36th Purdue Industrial Waste Conference*, Purdue University, West Lafayette, Indiana, pp. 860–866.

Metcalf and Eddy, Inc. 1993. *Wastewater Engineering: Treatment, Disposal, and Reuse*, New York: McGraw-Hill Book Co.

Morgan, T. W., et al. 1973. "Controlled Landfill and Leachate Treatment," Paper Presented at the *46th Water Pollution Control Federation Conference*, Cleveland, Ohio.

Nordstedt, R. A. 1975. "Land Disposal of Effluent from a Sanitary Landfill," *Journal Water Pollution Control Federation*, 47(7):1961–1970.

Palit, T. and S. R. Qasim. 1977. "Biological Treatment Kinetics of Landfill Leachate," *Journal of the Environmental Engineering Division*, ASCE, 103(EE2):353–366.

Pfeffer, J. T. 1986. "Treatment of Leachate from Landfill Disposal Facilities," *Proceedings of the Conference on Preparing Now for Tomorrow's Needs*, Chicago, Ill.

Pirbazari, M., B. N. Badriyha, V. Ravindran and S. H. Kim. 1989. "Landfills-B. Leachate," Section Fire, *Proceedings of the 44th Purdue Industrial Waste Conference*, Purdue University, West Lafayette, Indiana, pp. 555–563.

Pohland, F. G. 1972. "Landfill Stabilization with Leachate Recycle," Interim Progress Report, Grant EP 00658-01, Solid Waste Research Division, U.S. Environmental Protection Agency.

Pohland, F. G. 1975. "Sanitary Landfill Stabilization with Leachate Recycle and Residual Treatment," EPA 600/2-75-043, U.S. EPA, Cincinnati, Ohio.

Pohland, F. G. 1980. "Leachate Recycle as Landfill Management Option," *Journal of the Environmental Engineering Division*, ASCE, 106(EE6):1057–1069.

Pohland, F. G. and S. J. Kang. 1975. "Sanitary Landfill Stabilization with Leachate Recycle and Residual Treatment," *American Institute of Chemical Engineers*, Symposium Series No. 145, Water–1974, 71:308–318.

Pohland, F. G. and S. R. Harper. 1985. "Critical Review and Summary of Leachate and Gas Production from Landfills," Report to EPA, WERL, Coop. Agreement CR809997, Cincinnati, Ohio.

Pohland, F. G., M. Stratakis, W. H. Cross and S. F. Tyahla. 1990. Controlled Landfill Management of Municipal Solid Waste and Hazardous Wastes," *Proceedings of 1990 WPCF National Specialty Conference on Water Quality Management of Landfills*," Chicago, Ill., July 15–18, pp. (3-16)–(3-32).

Qasim, S. R., 1994. "Wastewater Treatment Plants: Planning, Design, and Operation," Lancaster, PA: Technomic Publishing Co., Inc.

Raina, S. and D. S. Malvinic. 1985. "Comparison of Fill and Draw Reactors for Treatment of Landfill Leachate," *Water Pollution Research Journal*, of Canada 20(2)12–28.

Robinson, H. D. and P. J. Maris. 1985. "The Treatment of Leachates from Domestic Waste in Landfill Sites," *Journal Water Pollution Control Federation*, 57(1):30–38.

Rogers, W. P. 1973. "Treatment of Leachate from a Sanitary Landfill by Lime Precipitation Followed by an Anaerobic Filter," thesis presented to Clarkston College of Technology, at Potsdam, New York, in partial fulfillment of the requirements for the degree of Master of Science.

Sanford, W. E., R. J. Kopka, T. J. Steenhuis, J. M. Surface and J. M. Lavine. 1990. "An Investigation into the Use of a Subsurface Flow Rock–Reed Filters for the Treatment of Leachate from a Solid Waste Landfill," *Proceedings of 1990 WPCF*

National Specialty Conference on Water Quality Management of Landfills," Chicago, Ill., July 15-18, pp. (4-1)-(4-18).

Schafer, P. E., J. L. Carter and G. C. Woelfel. 1986. "First Year's Operating Performance of the Omega Hills Landfill Pretreatment Anaerobic Filter," *Proceedings of the 41st Purdue Industrial Waste Conference,* Purdue University, West Lafayette, Indiana, pp. 383-389.

Simensen, T. and Odegaard, H. 1971. *Pilot Studies for the Chemical Coagulation of Leachate,* Oslo, Norway: Norwegian Institute of Water Pollution Research.

Slater, C. S., R. C. Ahlert and C. G. Uchrin. 1983. "Treatment of Landfill Leachate by Reverse Osmosis," *Environmental Progress,* 2(4):251-256.

Smith, C. M. and T. Krug. 1990. "Field Demonstration of Membrane Technology for Treatment of Landfill Leachate," *Ontario Ministry of the Environment Technology Transfer Conference,* Toronto, Canada, November 19-20.

Smith, R. G. and P. A. Wilderer. 1986. "Treatment of Hazardous Landfill Leachate Using Sequencing Batch Reactor," *Proceedings of the Purdue Industrial Waste Conference,* Purdue University, West Lafayette, Indiana, pp. 272-282.

Steiner, R. L., J. D. Keenan and A. A. Fungaroli. 1979. "Demonstrating Leachate Treatment Report on Full Scale Operating Plant," EPA-SW-758, Washington, D.C.: U.S. EPA.

Straub, W. A. 1980. "Development and Application of Models of Sanitary Landfill Leaching and Landfill Stabilization," RP No. 259, Resource Policy Center, Thayer School of Engineering, Dartmouth College, Hanover, New Hampshire.

Straub, W. A. and D. A. Lynch. 1982. "Models of Landfill and Leaching: Organic Strength," *Journal of Environmental Engineering Division,* ASCE, 108(EE2): 251-268.

Thornton, R. J. and F. C. Blanc. 1973. "Leachate Treatment by Coagulation and Precipitation," *Journal of the Environmental Engineering Division,* ASCE, 99(EE4): 535-539.

Tchobanoglous, G. and G. T. Schroeder. 1985. *Water Quality: Characteristics, Modeling and Modification,* Reading, MA: Addison-Wesley Publishing Co.

Tolman, A. L., A. P. Ballestero, W. W. Beck and G. H. Emrich. 1978. "Guidance Manual for Minimizing Pollution from Waste Disposal Sites," EPA-600/2-78-142, Cincinnati, Ohio: U.S. EPA.

Uloth, V. C. and D. S. Mavinic. 1977. "Aerobic Biotreatment of High Strength Leachate," *Journal of the Environmental Engineering Division,* ASCE, 103(EE6):647-661.

U.S. EPA 1979. "Process Design Manual for Sludge Treatment and Disposal," EPA-625/1-79-011, Technology Transfer.

Van Fleet, S. R., J. F. Judkins and F. J. Molz. 1974. "Discussion, Aerobic Biostabilization of Sanitary Landfill Leachate," *Journal Water Pollution Control Federation,* 46(11):2611-2612.

Vydra, O. M. and A. Grimm. 1976. "County Treats a Shredfill Leachate," *Civil Engineering,* ASCE, 46(12):55-57.

Whittaker, H., C. I. Adams, S. A. Salo and A. Morgan. 1975. "Reverse Osmosis at Gloucester Landfill," *Proceedings of the Technical Seminar on Chemical Spills,* Environment, Canada, pp. 190-207.

Wong, P. T. and D. S. Mavinic. 1984. "Treatment of a Municipal Leachate Under

Multi-Variable Conditions," *Journal Water Pollution Control Federation,* Research, Canada.

Zachopoulos, S. A., E. L. Tharp and P. C. Morgan. 1990. "Co-Disposal of Sanitary Landfill Leachate with Municipal Wastewater," *Proceedings of 1990 WPCF National Specialty Conference on Water Quality Management of Landfill,* Chicago, Ill., July 15–18, pp. (3-1)–(3-15).

Zapf-Gilje, R. and D. S. Mavinic. 1981. "Temperature Effects on Biostabilization of Leachate," *Journal of the Environmental Engineering Division,* ASCE, 107(EE4):653–663.

Soil Moisture Retention Tables Used for Calculation of Leachate Generation from Sanitary Landfills

Source: Adapted from Thornthwaite and Mather (1955 and 1957).

TABLE A.1. Soil Moisture Retention Table—100 MM.

SOIL MOISTURE RETAINED AFTER DIFFERENT AMOUNTS OF POTENTIAL EVAPOTRANSPIRATION HAVE OCCURRED. SOIL MOISTURE STORAGE AT FIELD CAPACITY IS 100 MM.

NEG (I-PET)	0	1	2	3	4	5	6	7	8	9
0	100	99	98	97	96	95	94	93	92	91
10	90	89	88	88	87	86	85	84	83	82
20	81	81	80	79	78	77	77	76	75	74
30	74	73	72	71	70	70	69	68	68	67
40	65	66	65	64	64	63	62	62	61	60
50	60	59	59	58	58	57	56	56	56	54
60	54	53	53	52	52	51	51	50	50	49
70	49	48	48	47	47	46	46	45	45	44
80	44	44	43	43	42	42	41	41	40	40
90	40	39	39	38	38	38	37	37	36	36
100	36	35	35	35	34	34	34	33	33	33
110	32	32	32	31	31	31	30	30	30	30
120	29	29	29	28	28	28	27	27	27	27
130	26	26	26	26	25	25	25	24	24	24
140	24	24	23	23	23	23	22	22	22	22
150	22	21	21	21	21	20	20	20	20	20
160	19	19	19	19	19	18	18	18	18	18
170	18	17	17	17	17	17	16	16	16	16
180	16	16	15	15	15	15	15	15	14	14
190	14	14	14	14	14	14	13	13	13	13

TABLE A.1. (continued).

SOIL MOISTURE RETAINED AFTER DIFFERENT AMOUNTS OF POTENTIAL EVAPOTRANSPIRATION HAVE OCCURRED. SOIL MOISTURE STORAGE AT FIELD CAPACITY IS 100 MM.

NEG (I-PET)	0	1	2	3	4	5	6	7	8	9
200	13	13	12	12	12	12	12	12	12	12
210	12	11	11	11	11	11	11	11	11	11
220	10	10	10	10	10	10	10	10	10	10
230	9	9	9	9	9	9	9	9	9	9
240	8	8	8	8	8	8	8	8	8	8
250	8	8	8	7	7	7	7	7	7	7
260	7	7	7	7	7	7	6	6	6	6
270	6	6	6	6	6	6	6	6	6	6
280	6	6	6	6	6	5	5	5	5	5
290	5	5	5	5	5	5	5	5	5	5
300	5	5	4	4	4	4	4	4	4	4
310	4	4	4	4	4	4	4	4	4	4
320	4	4	3	3	3	3	3	3	3	3
330	3	3	3	3	3	3	3	3	3	3
340	3	3	3	3	3	3	3	3	3	3

(continued)

TABLE A.1. (continued).

SOIL MOISTURE RETAINED AFTER DIFFERENT AMOUNTS OF POTENTIAL EVAPOTRANSPIRATION HAVE OCCURRED. SOIL MOISTURE STORAGE AT FIELD CAPACITY IS 100 MM.

NEG (I-PET)	0	1	2	3	4	5	6	7	8	9
350	3	3	3	3	3	3	3	3	3	2
360	2	2	2	2	2	2	2	2	2	2
370	2	2	2	2	2	2	2	2	2	2
380	2	2	2	2	2	2	2	2	2	2
390	2	2	2	2	2	2	2	2	2	2
400	2	2	2	2	2	2	2	2	2	2
410	2	2	2	2	2	1	1	1	1	1
420	1	1	1	1	1	1	1	1	1	1
430	1	1	1	1	1	1	1	1	1	1
440	1	1	1	1	1	1	1	1	1	1
450	1	1	1	1	1	1	1	1	1	1
460	1	1	1	1	1	1	1	1	1	1
470	1	1	1	1	1	1	1	1	1	1
480	1	1	1	1	1	1	1	1	1	1
490	1	1	1	1	1	1	1	1	1	1
500	1	1	1	1	1	1	1	1	1	1

TABLE A.2. Soil Moisture Retention Table—125 MM.

SOIL MOISTURE RETAINED AFTER DIFFERENT AMOUNTS OF POTENTIAL EVAPOTRANSPIRATION HAVE OCCURRED. SOIL MOISTURE STORAGE AT FIELD CAPACITY IS 125 MM.

NEG (I-PET)	0	1	2	3	4	5	6	7	8	9
0	125	124	123	122	121	120	119	118	117	116
10	115	114	113	112	111	110	109	108	107	106
20	106	105	104	103	102	102	101	100	99	99
30	98	97	95	95	94	94	93	92	91	90
40	90	89	88	87	86	86	85	84	84	83
50	83	82	82	81	80	80	79	79	78	77
60	76	76	75	74	74	73	73	72	72	71
70	70	70	69	69	68	68	67	67	66	65
80	65	64	64	63	63	62	62	61	61	60
90	60	59	59	58	58	57	57	56	56	55
100	55	55	54	54	53	53	53	52	54	51
110	51	51	50	50	49	49	49	48	48	47
120	47	47	46	46	45	45	45	44	44	43
130	43	43	42	42	41	41	41	41	40	40
140	40	40	39	39	39	38	38	38	38	37

(continued)

TABLE A.2. (continued).

SOIL MOISTURE RETAINED AFTER DIFFERENT AMOUNTS OF POTENTIAL EVAPOTRANSPIRATION
HAVE OCCURRED. SOIL MOISTURE STORAGE AT FIELD CAPACITY IS 125 MM.

NEG (I-PET)	0	1	2	3	4	5	6	7	8	9
150	37	37	36	36	36	35	35	35	35	34
160	34	34	33	33	33	32	32	32	32	31
170	31	31	31	30	30	30	30	30	30	29
180	29	29	29	29	28	28	28	27	27	27
190	26	26	26	26	26	25	25	25	25	25
200	24	24	24	24	24	23	23	23	23	23
210	22	22	22	22	22	22	22	21	21	21
220	21	21	21	21	20	20	20	20	20	20
230	19	19	19	19	19	18	18	18	18	18
240	18	18	17	17	17	17	17	17	17	17
250	16	16	16	16	16	16	16	16	16	15
260	15	15	15	15	15	14	14	14	14	14
270	14	14	14	14	14	13	13	13	13	13
280	13	13	13	13	13	12	12	12	12	12
290	12	12	12	12	12	11	11	11	11	11

TABLE A.2. (continued).

SOIL MOISTURE RETAINED AFTER DIFFERENT AMOUNTS OF POTENTIAL EVAPOTRANSPIRATION HAVE OCCURRED. SOIL MOISTURE STORAGE AT FIELD CAPACITY IS 125 MM.

NEG (I-PET)	0	1	2	3	4	5	6	7	8	9
300	11	11	11	11	11	10	10	10	10	10
310	10	10	10	10	10	10	10	10	9	9
320	9	9	9	9	9	9	9	9	9	9
330	8	8	8	8	8	8	8	8	8	8
340	8	8	8	8	8	7	7	7	7	7
	0	5		•	0	5	•	•	0	
350	7	7		450	3	3		550	1	
360	7	6		460	3	3		560	1	
370	6	6		470	3	3		570	1	
380	6	5		480	2	2		580	1	
390	5	5		490	2	2		590	1	
400	5	5		500	2	2		600	1	
410	4	4		510	2	2		610	1	
420	4	4		520	2	2		620	1	
430	4	4		530	2	2		630	1	
440	3	3		540	2	1		640	1	

315

TABLE A.3. Soil Moisture Retention Table—150 MM.

SOIL MOISTURE RETAINED AFTER DIFFERENT AMOUNTS OF POTENTIAL EVAPOTRANSPIRATION HAVE OCCURRED. SOIL MOISTURE STORAGE AT FIELD CAPACITY IS 150 MM.

NEG (I-PET)	0	1	2	3	4	5	6	7	8	9
0	150	149	148	147	146	145	144	143	142	141
10	140	139	138	137	136	135	134	133	132	131
20	131	130	129	128	127	127	126	125	124	123
30	122	122	121	120	119	118	117	115	115	114
40	114	113	113	112	111	111	110	109	108	107
50	107	106	106	105	104	103	103	102	101	100
60	100	99	98	97	97	97	96	95	94	93
70	93	92	92	91	90	90	89	89	88	87
80	87	85	86	85	84	84	84	83	83	82
90	82	81	81	80	79	79	78	77	77	76
100	76	76	75	75	74	74	73	72	72	71
110	71	71	70	70	69	69	68	68	67	67
120	66	66	65	65	65	64	64	63	63	62
130	62	62	61	61	60	60	60	59	59	58
140	58	58	57	57	56	56	55	55	54	54
150	54	53	53	53	52	52	52	52	51	51
160	51	51	50	50	50	49	49	48	48	47
170	47	47	47	46	46	46	45	45	45	44
180	44	44	44	43	43	43	42	42	42	41
190	41	41	41	40	40	40	40	39	39	39

TABLE A.3. (continued).

SOIL MOISTURE RETAINED AFTER DIFFERENT AMOUNTS OF POTENTIAL EVAPOTRANSPIRATION HAVE OCCURRED. SOIL MOISTURE STORAGE AT FIELD CAPACITY IS 150 MM.

NEG (I-PET)	0	1	2	3	4	5	6	7	8	9
200	39	38	38	38	37	37	37	37	36	36
210	36	36	35	35	35	35	35	34	34	34
220	34	34	33	33	33	33	33	32	32	32
230	32	31	31	31	31	31	30	30	30	30
240	30	29	29	29	29	29	28	28	28	28
250	28	27	27	27	27	27	26	26	26	26
260	26	26	25	25	25	25	25	24	24	24
270	24	24	24	23	23	23	23	23	23	23
280	22	22	22	22	22	22	22	22	21	21
290	21	21	21	20	20	20	20	20	20	20
300	20	19	19	19	19	19	19	19	18	18
310	18	18	18	18	18	18	18	17	17	17
320	17	17	17	17	17	17	17	16	16	16
330	16	16	16	16	16	16	16	15	15	15
340	15	15	15	15	15	15	14	14	14	14
350	14	14	14	14	14	14	14	13	13	13
360	13	13	13	13	13	13	13	12	12	12
370	12	12	12	12	12	12	12	12	11	11
380	11	11	11	11	11	11	11	11	11	11
390	11	11	11	10	10	10	10	10	10	10

(continued)

TABLE A.3. (continued).

SOIL MOISTURE RETAINED AFTER DIFFERENT AMOUNTS OF POTENTIAL EVAPOTRANSPIRATION HAVE OCCURRED. SOIL MOISTURE STORAGE AT FIELD CAPACITY IS 150 MM.

NEG (I-PET)	0	1	2	3	4	5	6	7	8	9
400	10	10	10	10	10	10	10	10	9	9
410	9	9	9	9	9	9	9	9	9	9
420	9	9	9	8	8	8	8	8	8	8
430	8	8	8	8	8	8	8	8	8	8
440	8	8	8	7	7	7	7	7	7	7
450	7	7	7	7	7	7	7	7	7	7
460	7	7	7	7	6	6	6	6	6	6
470	6	6	6	6	6	6	6	6	6	6
480	6	6	6	6	6	6	6	6	5	5
490	5	5	5	5	5	5	5	5	5	5
500	5	5	5	5	5	5	5	5	5	5
510	5	5	5	5	5	5	5	5	4	4
520	4	4	4	4	4	4	4	4	4	4
530	4	4	4	4	4	4	4	4	4	4
540	4	4	4	4	4	4	4	4	4	4
550	4	4	4	4	4	4	4	3	3	3
560	3	3	3	3	3	3	3	3	3	3
570	3	3	3	3	3	3	3	3	3	3
580	3	3	3	3	3	3	3	3	3	3
590	3	3	3	3	3	3	3	3	3	3

TABLE A.3. (continued).

SOIL MOISTURE RETAINED AFTER DIFFERENT AMOUNTS OF POTENTIAL EVAPOTRANSPIRATION HAVE OCCURRED. SOIL MOISTURE STORAGE AT FIELD CAPACITY IS 150 MM.

NEG (I-PET)	0	1	2	3	4	5	6	7	8	9
600	3	3	3	3	3	2	2	2	2	2
610	2	2	2	2	2	2	2	2	2	2
620	2	2	2	2	2	2	2	2	2	2
630	2	2	2	2	2	2	2	2	2	2
640	2	2	2	2	2	2	2	2	2	2
650	2	2	2	2	2	2	2	2	2	2
660	2	2	2	2	2	2	2	2	2	2
670	2	2	2	2	2	2	2	2	2	2
680	2	2	1	1	1	1	1	1	1	1
690	1	1	1	1	1	1	1	1	1	1
700	1	1	1	1	1	1	1	1	1	1
710	1	1	1	1	1	1	1	1	1	1
720	1	1	1	1	1	1	1	1	1	1
730	1	1	1	1	1	1	1	1	1	1
740	1	1	1	1	1	1	1	1	1	1
	0	5	•	•	0	5	•	•	0	5
750	1	1	1	750	1	1	1	830	1	1
760	1	1	1	800	1	1	1	840	1	1
770	1	1	1	810	1	1	1		1	1
780	1	1	1	820	1	1	1		1	1

REFERENCES

Thornthwaite, C. W. and J. R. Mather. 1955. "Instructions and Tables for Computing Potential Evapotranspiration and Water Balance," *Publications in Climatology,* Vol. 10, No. 3, Drexel Institute, Centerton, New Jersey, p. 86.

Thornthwaite, C. W. and J. R. Mather. 1957. "Instructions and Tables for Computing Potential Evapotranspiration and Water Balance," *Publications in Climatology,* Vol. 10, No. 3, Drexel Institute, Centerton, New Jersey, pp. 185–311.

List of Acronyms

BAC	biologically-active carbon
BOD	biochemical oxygen demand
CAA	Clean Air Act
CCI	construction cost index
CEC	cation exchange capacity
CERCLA	Comprehensive Environmental Response Compensation and Liabilities Act of 1980
Cl	chlorine
COD	chemical oxygen demand
CFR	Code of Federal Regulations
CPE	chlorinated polyethylene
CPE-R	reinforced chlorinated polyethylene
CQA	construction quality assurance
CQC	construction quality control
CR	neoprene (chloroprene rubber)
CSPE	chlorosulphonated polyethylene
CSPE-R	reinforced chlorosulphonated polyethylene
CSMoS	Center for Subsurface Modeling Support
ECO	epichlorohydrin rubber
EIA	ethylene interpolymer alloy
EIA-R	reinforced ethylene interpolymer alloy
ELPO	elasticized polyolefin
ENR	Engineering News Record
ENR-CCI	Engineering News Record-Construction Cost Index
EPDM	ethylene propylene diene mono rubber
FMLs	flexible membrane liners
FS	fixed solids
GLCs	geosynthetic clay liners
HDPE	High density polyethylene
HELP	Hydrologic Evaluation of Landfill Performance
HSWA	Hazardous and Solid Waste Amendments of 1984
LEL	lower explosive limit
LCRS	leachate collection and removal system

LDCRS	leachate detection, collection and removal system
LDPE	low density polyethylene LDPE
MACT	maximum available control technology
MCL	maximum contaminant level
MDPE	medium density polyethylene
MLE	modified Ludzack-Ettinger
MLSS	mixed liquor suspended solids
MLVSS	mixed liquor volatile suspended solids
MO	microorganisms
MSW	municipal solid waste
MSWLF	municipal solid waste landfill
MSWSLF	municipal solid waste sanitary landfill
NAAQS	National Ambient Air Quality Standards
NGWIC	National Ground Water Information Center
NSWMA	National Solid Waste Management Association
NPDES	National Pollution Discharge Elimination System
ORP	oxidation reduction potential
POTW	publicly owned treatment works
PVC	polyvinyl chloride
PVC-E	elasticized polyvinyl chloride
PSD	prevention of significant deterioration
RBC	rotating biological contactor
RCRA	Resource Conservation and Recovery Act
RSKERL	Robert S. Kerr Environmental Research Laboratory
SARA	Superfund Amendments and Reauthorization Act of 1986
SBR	sequencing batch reactor
SDWA	Safe Drinking Water Act
SIP	State Implementation Plan
SLF	sanitary landfill
SML	Synthetic membrane liner
SO4	sulfate
TCLP	Toxicity Characteristics Leaching Procedure
TDS	total dissolved solids
TOC	total organic carbon
TPE	thermoplastic elastomer
TSCA	Toxic Substance Control Act of 1976
USDAHL	United States Department of Agriculture Hydrograph Laboratory
U.S. EPA	United States Environmental Protection Agency
VLDPE	very low density polyethylene
VOC	volatile organic carbon
VS	volatile solids

Index